The Wooster Book Company
205 West Liberty Street
Wooster, Ohio 44691
www.woosterbook.com

Printed in the United States of America.

ISBN: 978-1-59098-485-7

Library of Congress Cataloging-in-Publication Data
Minnick, Fred, (date)
The brand that changed beef : how the Certified Angus Beef brand became a worldwide icon of quality : a company biography / Fred Minnick.
p. cm.
ISBN 978-1-59098-485-7 (alk. paper)
1. Beef—Marketing. 2. Brand name products. 3. American Angus Association. Certified Angus Beef. I. American Angus Association. Certified Angus Beef. II. Title.
HD9433.A2M56 2010
381'.416223—dc22 2010001886

THE BRAND THAT CHANGED BEEF

How the *Certified Angus Beef*® Brand Became a Worldwide Icon of Quality

THE BRAND THAT CHANGED BEEF

How the *Certified Angus Beef* ® Brand Became a Worldwide Icon of Quality

A Company Biography

Fred Minnick

The Wooster Book Company
where minds and imaginations meet

The Wooster Book Company
Wooster, Ohio
2010

Product. People. Integrity.

The Certified Angus Beef brand's success is shaped by its product quality, integrity, and the people who indelibly show a genuine passion for the brand. This book is dedicated to each person who was a part of shaping its incredible history and to those carrying the torch for its future. To the licensees who believed in the brand enough to stake their own business reputation on it—we thank you. To the American Angus Association members—it's with great pride that we take your brand to the marketplace each day. And, to the individual staff members who put their heart and soul into every aspect of their jobs—thank you for your part in the brand's success.

This book outlines several key moments in the brand's storied history, but not all. Many, many more current and former staff, licensees, board members, cattlemen, and consultants have left their mark on the brand and led its success. While not all could be interviewed for this book, their contributions are greatly appreciated.

Just as the early brand pioneers showed great vision, this book would not be possible without the incredible leadership of two staff members—Crystal Meier and Steve Suther. Crystal started working on this book nineteen years ago, her first day on the job in our marketing department. Crystal's knowledge of our brand's history and records is incredible. Without a doubt, her personal commitment to excellence, attention to detail, and persistence is evident on every page. Steve Suther has told the brand's story through the written word to cattlemen for years. He helped shape the early vision for the book and has pored over each word, paragraph, and chapter with the bar held high for perfection. Thank you, Crystal and Steve, for your personal commitment to an accurate account of our history but, more importantly, for capturing the spirit and flavor of this great brand.

Thanks to all, from producer to consumer.

John F. Stika

John F. Stika, Ph.D.
President
Certified Angus Beef LLC

Author's Acknowledgements

I am so honored to have written this iconic beef brand's incredible history. But I certainly could not have done it alone.

Steve Suther recommended me as a candidate after reading a handful of my *Angus Journal* stories. Steve improved the manuscript with his great editorial eye and showed the many possibilities this brand creates for Angus producers. We spent a good deal of time together touring feedlots and packing facilities. Also an Angus producer and fantastic writer, Steve is an incredible talent.

Serving as the book's project manager, Crystal Meier could have written this company history in her sleep. Instead, I had the great fortune to work with her for more than a year. If I had a question about a date or fact, Crystal usually had an answer within thirty minutes. She served as first editor, coordinated my travel, set up many interviews, and became a good friend in the process. If it were not for her passion and foresight, this book would have never happened.

At the Certified Angus Beef office, so many people helped me find important materials and showed me around the area. Thank you all for sharing your perspectives and helping me find my way when I was lost in Wooster. I'd like to acknowledge Al Kober, whose intelligence and character shaped many of my personal opinions, helped me write the retail sections of this book, and helped make me a better man. His legacy of unwavering faith, integrity, and passion for the brand remains a guiding beacon for all.

I interviewed so many outstanding Certified Angus Beef licensed partners. I'd like to thank the following for showing me their facilities: Dole & Bailey, DeBragga and Spitler, Oxford Trading Company, Tyson Fresh Meats, Giant Eagle, Mackay Restaurant Group, MacDonald Meat Company, Summitcrest Farms, Henry Bergfeld's Pine Hill Farm, Buehler Food Markets, King's Command Foods, West Concord Super Market, Roche Bros., Café Escadrille, Waldorf=Astoria, Gregory Feedlot, Rosario's Grille, Smithco Meats, and Quality Food Centers.

I'd like to specifically thank Kris Black, of MacDonald Meat Company, for filling in at the last second, and then answering my questions as if she had months to prepare. John Stowell, of Dole & Bailey, opened his brilliant beef mind and shared so much invaluable input. He also treated me to one of the best-prepared Certified Angus Beef steaks at Café Escadrille. Charlie Robinson, of Oxford Trading Company, genuinely believes in this program and his tremendous enthusiasm rubbed off on me. Eliott Reyes Sr., of Provimi de Puerto Rico, explained over the phone his passion for Certified Angus Beef in a way that made me feel like I was in the room.

Charles Cannon, an original committee member, invited me into his Kentucky home and talked over a cup of coffee about how this brand changed the breed. His perspectives helped me shape the early chapters. And former president Jim Riemann passionately described his tenure and thankfulness for being a part of this company. He helped me understand the depths of the international market closures and the pride everybody took in the 2002 Olympic Winter Games. I also personally grew from interviewing Jim, by gaining insight into his family values and business principles.

The American Angus Association played a significant role in my research process. Keith Evans, former public relations director for the Association and author of *A Historic Angus Journey*, spent a day with me digging through old files, looking at pictures, and explaining details about Certified Angus Beef's beginnings. Sheila Stannard, Association director of activities and events, kindly helped me search through archived photos. I'd also like to thank CEO Bryce Schumann for an incredible interview. His appreciation for Certified Angus Beef and its history was evident in our talks. Much of my research would have hit a dead-end if it were not for the online *Angus Journal* archives. This great magazine chronicled every step of Certified Angus Beef's life and stored it all in searchable form on *AngusJournal.com*.

I'd like to thank Dr. Bob VanStavern and Mick Colvin. You two gentlemen helped me travel back in time to when you fought to keep this brand alive. Interviewing you and reviewing past interviews with Fred Johnson helped me realize that Certified Angus Beef is not just about a product or a breed of cattle. It has and always will be about great people with the utmost integrity.

To the entire staff and company, thank you for letting me tell your history.

Fred Minnick

Table of Contents

John Q's Steakhouse in Cleveland is one of thousands of restaurants that require the Certified Angus Beef brand's superior quality. Every night at John Q's, finely dressed waiters and waitresses show guests a tray of steaks and explain the unique flavor characteristics of each.

The Big Idea

As diners take their seats inside John Q's Steakhouse, downtown Cleveland, a nicely dressed waiter shows them a tray of thick, well-marbled steaks. Before the waiter can describe the cuts, a man points at the bright-red filet in the middle. "What's that one?" he asks with anticipation.

"I see you have good taste," the waiter responds. "This is our center-cut filet mignon and like all of our steaks, it is the Certified Angus Beef brand, which is the best beef in the world."

He goes on to describe the superior attributes of the New York strip steak, the Delmonico, and the ten-ounce top sirloin steak topped with melted bleu cheese and a red wine demi-glace. As the waiter gives his convincing spiel, the guests admire John Q's menu. On the right-hand side is a box with the brand's logo and a paragraph of high praise: "John Q's proudly serves only *Certified Angus Beef®*, the highest quality beef available. Only the top of the USDA Prime and Choice grades are labeled and sold under the *Certified Angus Beef®* trademark."

At another table, the waiter again describes how this beef is better than anything found at the steakhouse's competitors. Some people might think those are just words from a waiter trying to increase the check for a larger tip. But if they press their fork on the tender bone-in rib steak and watch the natural beef juices flow, they check their cynicism at the door and bite into the best, most consistent beef in the world.

John Q's is one of thousands of Certified Angus Beef licensees around the world proudly serving the brand. From fine-dining restaurants like the Bull and Bear steak-

At El Gaucho steakhouses, guests receive personal attention with tableside preparation of Caesar salads and carving of their Chateaubriand. The fine-dining restaurant is considered a steakhouse legacy in the Northwest
Photo credit:Ilya Photography for El Gaucho

Opposite: Licensed restaurants proudly display the Certified Angus Beef logo and explain why only the top USDA Choice and Prime beef qualify.

house at the Waldorf=Astoria in New York, which serves celebrities on a daily basis, and El Gaucho in Seattle, where large parties will spend thousands, to the Taste of Texas in Houston, where warm hospitality meets tradition; and major grocery store chains like Meijer in Grand Rapids, Michigan, Quality Food Centers in Seattle, and Price Chopper in Schenectady, New York, food companies understand the brand's name signifies superior quality, consumer awareness, and proven sales of some $2.7 billion per year. In 2009, consumers purchased nearly two million pounds per day of product. That comes to an estimated 3.5 billion (three-ounce) servings a year from 15,100 business partners. This popularity has spawned generic "Angus" competitors, which sell lower quality beef. At last count, there were more than sixty other Angus programs. But the Certified Angus Beef brand is the only Angus program owned by the American Angus Association®, and every pound is tracked.

There's truly no other branded meat product in the world like this one. The Certified Angus Beef logo has become an icon for a brand that boosts revenues for everybody involved. Licensed packing companies pay Angus producers an estimated $25 million in annual premiums to encourage more supply. That has led more producers to change to Angus.

In 2009, more than 60 percent of U.S. cattle were black hided, with Angus bulls selling at healthy premiums, and the brand's logo commanding consumer and breeder awareness around the world. It was marketed to consumers in forty-seven countries, in more than eight thousand restaurants and more than fifty-six hundred grocery stores.

However, this is not your typical company success story. Unlike Google, McDonald's, or Microsoft, the founders of this brand did not become filthy rich. According to the people who started the not-for-profit company in 1978, the program's objective was and always will be to improve demand for better Angus cattle, not to make millions in profits.

John Q's
STEAKHOUSE
Established 1979
APPETIZERS
French Onion Soup 6
Traditional French onion soup served with a parmesan crouton and topped with Gruyere Swiss.
Jumbo Shrimp Cocktail 11
Garlic Shrimp 11
Four jumbo shrimp sautéed with white wine, garlic, tomatoes, capers and lemon.
Stuffed Banana Peppers 10
Banana peppers stuffed with chorizo sausage, ground beef and rice, served with marinara sauce and provolone cheese.
Seared Sea Scallops 12
Served with Cajun cream sauce.
Cheese-Stuffed Mushrooms 8
Baked mushroom caps stuffed with four cheeses, garlic and butter.
1/2 lb. Alaskan King Crab Legs 15
Pita Chips with Artichoke Dip 9
Fresh pita chips lightly fried and topped with provolone cheese, scallions and tomatoes, served with a creamy artichoke dip.
Beer-Battered Onion Rings 7
Fried Calamari with Marinara Sauce 10
Crab Cake 10
Served with roasted red pepper aioli.
STEAKS & CHOPS
All entrées served with a Caesar salad and fresh vegetable or potato.
John Q's proudly serves only "Certified Angus Beef" the highest quality beef available. Only the top of the USDA prime and choice grades are labeled and sold under the Certified Angus Beef trademark.
12 oz. New York Strip Steak 31 16 oz. 38
12 oz. Black Pepper Strip Steak 32
Pan-fried with a coating of cognac mustard sauce and cracked black pepper.
8 oz. Center-Cut Filet Mignon 30 12 oz. Bone-in 39
24 oz. Porterhouse Steak 43
18 oz. Bone-in Rib Eye Steak 38
14 oz. Delmonico Steak 29
10 oz. Top Sirloin Steak 23
Grilled and topped with melted bleu cheese, served with a red wine demi-glace.
Slow-Roasted Prime Rib of Beef
(limited quantities nightly)
12 oz. 28 16 oz. 31
16 oz. Two Center-Cut Pork Chops 21
14 oz. Rack of Lamb 29
ADDITIONAL SELECTIONS 5
Fresh Vegetable • Baked Potato • Steak Fries • Q Chips • Potato of the Day
Wild Rice Pilaf • Sautéed Mushrooms • Grilled Onions
Add to any Steak – Garlic Shrimp 11 · Lobster Tail 25
SAUCES 2.50
Béarnaise • Cognac Mustard
Bleu Cheese with Red Wine Demi-Glace
Caramelized Onion Demi-Glace
SEASONINGS 1.50
Blackened • Cracked Black Pepper • Garlic
JOHN Q'S STEAK TEMPERATURE CHART
Charred Rare:
Rare: Cool center, red from edge to
Medium Rare: Warm red center
Medium: Pink throughout
Medium Well: Slightly pink
Well Done: No pink, dry
At John Q's we pride ourselves on our food and service. Your satisfaction is our primary
your business and if I can be of any assistance, please ask for me.
Orlando's Famous Fresh Ciabatta Bread
Gift Cards Available
Private Party Facilities
All Items and Prices are Subject to Change

Licensed Certified Angus Beef retailers display the brand's marketing materials and label every package. In addition to providing marketing materials, Certified Angus Beef conducts training seminars to help meat associates understand the product's science-based quality standards.

The organization faced numerous tough battles just to survive. A government agency doubted its legitimacy, a parent company was ready to pull the plug, and a mountain of skeptical packers as well as distributors and restaurateurs just didn't understand the concept. And to top it off, the 1970s had seen a growth in the popularity of Continental cattle breeds such as Limousin, Charolais, and Simmental, which generally produced lower marbling beef. For an industry that did not understand consumer demand, commodity beef was king, and the "war on fat" was top of mind for consumers.

Today, it's hard to imagine the beef industry without the Certified Angus Beef brand. But back then, their theories, ideas, and methodologies were so alien that they were truly rogues of the premium nature in a commodity world.

"It is trusted by producers and consumers alike," says Randy Blach, CEO of CattleFax, perhaps the best known independent research firm specializing in the cattle industry. "The integrity behind their brand has really been the key. The Certified Angus Beef

In Oregon, the Thomas family—Rob, Lori, Ty, and Bryson—takes great pride in raising the best Angus beef for consumers. As members of the American Angus Association, the Thomas's are among the cattlemen and women who are dedicated to producing the best. Certified Angus Beef is the only Angus brand owned by the 30,000-member Association.

brand was the first program that gave cattlemen the opportunity to focus on grade, to concentrate on differentiating product line, and to get paid for producing quality."

In many respects, the brand changed the beef industry. Craig Huffhines, CEO of the Certified Hereford Beef brand, founded in 1995, says the original Angus brand laid the groundwork for the U.S. Department of Agriculture's (USDA) specification system.

"It was the first real branded beef icon that launched a proliferation of branded concepts," says Huffhines, whose company competes with Certified Angus Beef. "Because they were able to make it successful and break the commodity barrier, it really encouraged many of us to put our first foot forward."

What would the beef industry look like if the Certified Angus Beef brand had never existed? As consumers, we may have had to settle for less flavor and consistency; we may have turned away from beef. And as cattlemen, we might never have launched the widespread branded beef movement that helped restore consumer confidence in beef. Thank goodness that question is hypothetical.

Luckily for today's beef consumers and Angus breeders, company founders were fighters hungry for a challenge. Despite all odds, they did not give up.

Looking back, those who made it happen often remarked, "We didn't know it couldn't be done."

SINCE 1978
CERTIFIED ANGUS BEEF®
BRAND
Angus beef at its best
CERTIFIED ANGUS BEEF® Brand

There's a Lot to the Name

As you read this book, you might notice the trademarked three-word name is used every time, rather than just one or two of the words or an acronym. While people may occasionally reference the brand in some other way, it is important to note why the Angus producers who own the company want the brand name spelled out.

Over the years, several trademarks have become generic terms: kerosene, aspirin, and dry ice; Kleenex, Band-Aid, and Rollerblade struggled to maintain their trademarks, and in England, Hoover refers to all vacuum cleaners. When brands fail to protect their trademarks, their names fall into the deep, dangerous abyss of brand generics among consumers.

Furthermore, there are companies that use acronyms for their names and attempt to build equity under the condensed version. How many Americans know what AT&T or IBM stand for? This branding strategy, too, has its dangers.

Unlike many companies, this one sees great opportunity in the *Certified Angus Beef®* brand name. Because of its popularity over the years, countless imitator brands have come and gone, trying to capitalize on the power of the "Angus" name. But the word "Angus" doesn't mean the Certified Angus Beef brand, and that three-word phrase is not a term that means a cut of meat is "certified" to be "Angus." It is simply the most recognized brand of fresh beef in the world, the only one owned by the Association of Angus producer-members, with every pound tracked from USDA certification to consumer purchase.

At the same time, you may notice in this book that the coveted ® that denotes the brand's registered name is not used every time. This change was made to make the text easier to read as the name is used numerous times throughout the copy. And the company certainly does not want to abbreviate and risk consumers confusing the pronunciation of its acronym with a yellow car in New York City—that would be a branding nightmare. As you will learn in Chapter 5, "Protecting the Brand," this company is diligent about protecting its name, trademarks, and people.

Mick Colvin (right) shakes the hand of the meat manager at Renzetti's IGA in Columbus, Ohio, where the first pound of Certified Angus Beef product is sold on October 18, 1978.

1 The Beginning

To be an Angus producer in the early to mid 1970s was disheartening. Continental breeds were growing and registrations at the American Angus Association were dropping at an alarming rate, while demand for Angus bulls was plummeting.

"As we continued to introduce more Continental breeds that did not have the capacity or capability of achieving a Choice-or-higher quality grade, we had less high-quality product on the market," remembers CattleFax CEO Randy Blach. "The industry became focused on growth and less on quality."

During the 1970s, as in any current decade, most people could not see the trends that were being established. Diet and health professionals recommended leaner beef, but few seemed to realize the difference between waste fat and taste fat. The cattle industry simply turned to the Continental cattle because of their rapid growth, but they produced beef with less flavor. Feeding them longer only produced waste. It was definitely an era of poor-quality beef, when even meat scientists supported the move to fewer flecks of white taste-fat in the middle.

"Beef had lost its exclusiveness," says Bobby "Dr. Bob" VanStavern, meat scientist and professor emeritus at The Ohio State University. "You could try to eat a steak, but you may or may not be able to chew it. You could pay a good price in a restaurant and have to return the steak. It was getting to the point where ground beef was about the only thing consumers could depend on. The industry was just about ready to turn into a ground beef industry. Beef did not have any other quality characteristics that we bragged about. We'd lost it. And producers were riding right along, because there

First imported to the United States from Aberdeen, Scotland, in 1873 by London businessman George Grant, Angus cattle have become synonymous with quality. They consistently grade better, and their beef is often more tender than that from other cattle breeds. Founded in 1883, the American Angus Association has always taken a scientific approach to improving herds and maintains a strict focus on improving beef quality.

wasn't any pricing mechanism to differentiate quality beef. It was becoming a lower-quality commodity."

So, how in the world could a premium beef idea come to fruition in this yield-driven market that seemed headed toward a ground beef future? As VanStavern says, cattlemen were not being financially rewarded for growing a premium product and many signs pointed toward beef becoming the next poultry. Ranchers were advised to forget flavor and cut costs to compete with retail chicken prices. But as producers tried to create more pounds of commodity beef, restaurateurs and butchers were starving for quality. There was a demand. The industry as a whole just didn't see it.

"There was confusion created with the war on fat," Blach says. "A lot of people took that to mean the consumer didn't want any marbling."

Harold Etling was not one of those people. A successful Angus breeder since 1944 at Wayne Knolls Farm in Marshallville, Ohio, and an astute businessman featured in *The Wall Street Journal*, Etling knew his breed of cattle could produce superior meat and was tired of the commodity market driving down his bull premiums. Etling had just about had it, believing something had to be done to save the Angus name. But what?

All it took was one more disappointing cut of beef and the wheels were turning. One evening in early 1975, Etling sat down to dinner in a high-class restaurant and returned a bad steak. What made it especially bad was the Angus label on the menu. When he chewed the hardly edible steak, Etling knew such affiliations with unpredictable and unmonitored quality would put Angus producers out of business.

Soon thereafter, Etling discussed the matter with American Angus Association Board member and friend Fred Johnson.

"Harold said we should do something about promoting Angus beef," Johnson recalled in a 1999 interview.

The two determined there was an opportunity to sell real Angus products to restaurants. They decided to bring in Elliot Frank, retired senior vice president of the LaSalle Bank and former chairman of the Illinois State Toll Highway Authority. Frank had worked with Marriott to develop the Great America Theme Park in Gurnee, Illinois, and was good friends with several Marriott executives. Moreover, he was a partner in Picket Fence Farm of Grayslake, Illinois, with Chicagoland real estate developer A. Harold Anderson. It appeared that Frank and his connections could be instrumental in developing some kind of Angus beef program with Marriott.

Through letters and phone calls, the trio discussed the issue. Frank knew such a program had to start with Marriott because of his close ties and the hotel chain's desire for high-quality beef.

In April 1975, the first in-person gathering took place at Picket Fence Farm. Attending the unofficial meeting were Johnson, Anderson, Frank, Robert Barrie—a Marriott official who helped decorate the Black Angus Room in the Robidoux Hotel at St. Joseph, Missouri—and Jerry Helgren, manager of Picket Fence.

"We were discussing how to get better beef to Marriott," Helgren says. "After that meeting, Elliott said, 'Well, it's a bigger thing than what one farm could do. We should get the American Angus Association involved.' "

Frank and Johnson spearheaded efforts to set up a meeting between Association officials and Marriott executives. Meanwhile, Etling investigated restaurants claiming to sell "Black Angus Steaks." It turned out that the tough steak Etling tried to eat probably didn't even originate from a black-hided animal.

"In checking with some of these Black Angus steakhouses, I found none used Angus exclusively, and that the Angus used was not very top quality," Etling wrote in an August 1975 letter, now considered the original blueprint for Certified Angus Beef. The supplier to one of those restaurants said he furnished them with beef; "some of it was Angus—some was other beef and not all really top quality." Etling said the supplier compared the Angus situation to selling rabbit sausage made with one rabbit and one horse, because there was no consistency with the quality of so-called Angus beef and anything could be called Angus. Etling continued to urge Johnson and the Association to establish a certification or franchise program for restaurants and meat shops to sell only top-quality Angus meats. That would protect the Angus name and ensure consumers get a premium product, he said.

"This program would require a rigid contract and inspection program on the part of the Association," Etling wrote. "Before a franchise or certification were granted, it would be necessary to have the source of supply thoroughly checked out as to their ability to supply Prime and high Choice Angus of the right grading, size, and cutability. The contract with the supplier would have to be approved and a license franchise or certification would have to be predicated on the reliability and credibility of the applicant."

In October 1975, Etling's idea was presented to the Association's board and Marriott officials, including renowned meat consultant Leonard Berkowitz, Barrie, and R.W. Marriott Jr. The meeting was a success. Marriott loved the idea and according to a December 1975 *Angus Journal* editorial by Jim Orton, the hotel chain planned to sell high-quality Angus cuts. But there were no systems in place to make sure they indeed received and sold Angus beef. Despite the enthusiasm from both parties, nothing really came from the Marriott meeting.

Nonetheless, the Association saw great potential in a program promoting high-quality Angus beef. At the October 1975 meeting, Association president Bill Clark said he would select an advisory committee to look into the program. Adding urgency to the matter, USDA grading standards were changed on February 23, 1976, relaxing maturity and marbling specifications for Choice, and maturity standards for Prime beef. Now, all beef in the upper two-thirds of Choice would be lumped in with a new, lower Choice portion that was called "Good" prior to the change. But consumers would never know the difference just by looking at the packaged meat. Johnson knew the Association had to look into Etling's idea. And soon.

"The Choice grade of beef was so broad that about 85 percent of beef graded fell into that category," Johnson said years later. "The program we developed took that into account, and the conditions set it up for us."

On March 13, 1976, the first "Angus Beef Certification"—the name would later be changed—Advisory Committee met in St. Joseph, Missouri, at the Ramada Inn. Chaired by Gilman Stewart, the committee included Etling, Johnson, Edward Elliot, Bob Geiss,

1975

Harold Etling (pictured) contacts Fred Johnson about "franchise" for Angus beef

Johnson meets with Picket Fence Farm representatives and Robert Barrie from Marriott

After a subsequent Marriott meeting, Bob Geiss presents Johnson's proposal to American Angus Association Board

Angus Beef Certification Committee formed

At Picket Fence Farms in Grayslake, Illinois, are (left to right) co-owner Elliot Frank, manager Jerry Helgren, and co-owner A. Harold Anderson. In April 1975, the first unofficial Certified Angus Beef meeting takes place here. Attending are Frank, a retired senior vice president of LaSalle Bank and frequent Illinois political appointee; Anderson, a Chicago real estate mogul; Helgren; Fred Johnson; and Robert Barrie, a Marriott official known to Frank and Anderson.

and Keith Evans. Also in attendance at the first meeting were Frank and Association officers Lloyd Miller, Roy Meier, Charles Cannon, Bill Bradley, John Smith, Joe Freeman, and Eli Votaw.

They all agreed that the objective was to devise a program to promote the sale and acceptance of Angus beef to consumers. They determined that about 85 percent of restaurants claiming to serve Angus were serving an inferior product but there was a strong demand to offer quality. They scratched the surface on other important issues, such as supply, public relations, marketing, inspection systems, and the possibilities of franchising restaurants.

"Following the lengthy committee discussion concerning the production, promotion, and marketing of end-product beef—like Black Angus—it is my opinion that a worthwhile consumer-marketing program can and should be developed, implemented, and promoted to benefit the Angus breed," wrote Elliot, who raised Angus cattle and was the first farmer to serve as president of the Ohio Restaurant Association. "Such a program should also certify the various delineated businesses that procure, market, and serve only top-quality Angus beef. Breeders have expressed the opinion that America's leading beef-breed registry association should be given all the support possible in marketing and promotion of their finished product if this breed is to continue to expand."

According to Evans' *A Historic Angus Journey: The American Angus Association 1883–2000*, the board tabled the program until funds were available. But the special committee unofficially met throughout 1976, and Louis "Mick" Colvin, an Ohio-based field representative for the Association, replaced Evans as the staff contact. The committee reported to the board in 1977 and recommended officially researching a certification program. Revived, the Angus Beef Certification Committee met with the board again. Its requests for legal counsel, industry-specific assistance, and a director were approved in November, along with $50,000 in expenses for the 1978 fiscal year. Etling's dream was on its way to becoming a reality: The idea was firm, the people in place were solid, and the Association supported the bold new venture.

1976

First Angus Beef Certification Committee meeting held; chairman Gilman Stewart (pictured), Harold Etling, Edward Elliott, Bob Geiss, Fred Johnson, and Association staff member Keith Evans

Committee meets with Marriott executives in Washington D.C.

Due to lack of funding, Board tables program; committee members continue meeting with packers and retailers

The Execution

A great idea doesn't mean much if the right people cannot execute a plan. Etling's letters to the Association encouraged it to create a franchise or licensing program to brand "specifically qualified Angus," but he admitted not having the stamina to lead such a program.

"What I am suggesting is not for my own interests as I am now seventy-five years of age and I probably will not want to continue in the meat and cattle business too many more years, but I think it is a shame to let a great breed of cattle suffer at the hands of the inferior breeds that are currently being promoted," Etling wrote. "If some of the directors are apathetic or want to run scared, there are really no insurmountable problems. All it would take is enthusiastic effort behind it and some work on the part of the Association to make this a great thing for our breed."

The Angus breed would not "run scared;" rather, it faced the challenges head on. The right people in the right place met this idea at the right time. And like Etling, they were not in it for selfish reasons.

"We all just wanted to help the breed," remembers Colvin, the program's first executive director. Prior to that post, he had considerable experience with the Association and as an Angus herd manager at Mole's Hill Farm in Connecticut and Longleaf Plantation in South Carolina. Colvin was a lanky young man who loved to learn, had a passion for Angus cattle, and seemingly never faced a challenge he couldn't handle.

A committee of great men, each with a long list of amazing accomplishments, met Colvin's enthusiasm. They were charged with organizing and executing the Angus Beef Certification Program. Charles Cannon was a board member and owner of Kentucky's

Above: The committee also includes Charles Cannon, a Kentucky Angus producer and Association board member, who later serves as board president in 1982.

Right: The Angus Beef Certification Program begins January 1, 1978. Committee members leading its development are (front row, from left to right): Mick Colvin, executive director, and Harold Etling; back row: consultant Dr. Bob VanStavern, Elliott Frank, Edward Elliot, and Fred Johnson.

Stonegate Farms, where he managed some two hundred registered brood cows. Elliott served Angus beef from his farm at his restaurant and was a national leader in the restaurant industry. Etling, the letter writer, was the founder of Weatherseal Inc., former president of the Ohio Angus Association, and part-owner of a packing plant. Frank was a career banker and political appointee to various Illinois positions. Committee chairman Fred Johnson, the co-owner of Summitville Tiles Inc., invented unique woodgrain ceramic tiles still used in fast-food restaurants around the world, and he was a nationally renowned Angus breeder as owner of Summitcrest Farms.

"We had some really successful, knowledgeable people to start," says Cannon. "I was thirty-six when I went on the board and Etling and all those guys were close to twice my age and successful businessmen. They had so much experience, not only in sales of beef but in tile, windows, and banking. Working with those guys was the highlight of my life in the Angus business."

It was definitely a "Who's Who" in business. Every one of them managed multimillion-dollar farms or major national corporations. But there was an obvious hole in the list. Elliot knew the restaurant business, Frank understood the financial world, Etling and Johnson were solid entrepreneurs, Colvin was an exceptional organizer, and they all knew the Angus business. But nobody was really a meat-science expert. That was an area in which they could not afford novice recommendations.

Colvin knew the yet-unwritten scientific specifications for the program would be absolutely paramount to its success. The director reached out to VanStavern, who was constantly getting inquiries from packers and restaurateurs about the lack of quality beef in the market.

"This program was the epitome of everything I had ever tried to do with the industry starting with production and seeing it right through to the consumer," says VanStavern, who earned the 1973 American Meat Science Association Distinguished Extension

Frank Murphy's paintings forever capture the Angus breed's subtle curves and rich black hide in natural settings. The American Angus Association frequently commissions Murphy to paint Angus cattle and has used his work in brochures and advertisements for more than fifty years. This steer becomes the basis for the beef brand's logo. Courtesy of American Angus Association and Frank C. Murphy

Phase I and Phase II

The original committee envisioned two phases to the program.

Phase I, which continues today, involved maintaining identity of Angus-type cattle through processing in licensed packing plants. The grader under USDA's Beef Evaluation Program certified only Angus carcasses that met or exceeded the standards. They were then marked with the Certified Angus Beef logo roll, the device graders used to designate the cattle. Then the beef was merchandised in a section of a licensed retailer's beef case, clearly identified as the brand.

Phase II was to go back to the farm to gather data on Angus cattle sired by various registered Angus bulls. With the use of numbered ear tags and computers, the plan was to return information, perhaps even premiums, to producers for each calf that qualified.

Mick Colvin (right), regional manager for the American Angus Association, visits with Angus producer Ed Murphy at an Ohio Angus Field Day. Thought to be one of the great young minds in the cattle business, Colvin is selected by the committee to develop the brand. Colvin retires twenty-two years later, after leading the brand from the first pound sold to substantial international sales and recognition.

Industry Service Award. "All these restaurant and distributor groups had been telling me their problems for years."

In 1977, VanStavern was one of the few meat scientists in the country not swayed by the commodity-beef trend. While others in his profession were leaning toward yield standards, VanStavern stuck with quality. He still judged based on carcass merit and preferred young cattle that produced at least moderate marbling—traits that led to premium beef and favored Angus cattle.

He jokingly says it took him "about thirty seconds" to come up with the original specifications for the Angus Beef Certification Program. Those included, after live-animal screening to evaluate only cattle with 51 percent or more black hide: Modest or higher marbling, "A" maturity on a scale of A–E, medium to fine marbling texture, Yield Grade 3 or leaner cutability, no dark cutters, and no blood spots. VanStavern and Colvin held firm when a packer suggested both A and B maturity. The specifications have not changed appreciably over time, except for clarifications on how to exclude cattle with Brahman or dairy heritage and more detailed parameters that replaced yield grade and enhanced uniformity.

While the criteria were well accepted, some producers were displeased with the black-hided percentage, fearing it would allow non-Angus into the program. But the Association Board realized that most cattle raised for meat, rather than breeding, were not purebred. The black hide in combination with all the carcass specs would ensure primarily Angus heritage and form the basis for a market-driven supply. In other words, producers who wished to qualify more cattle for the brand would raise higher-percentage Angus cattle of a certain type.

However, quality and cattle type were not the only issues at this stage.

"I had been involved with some pork programs previously, based on specifications that were to result in a greater value product, and those programs failed because we did not have third-party evaluation," VanStavern says. "We needed USDA for two reasons: certifying the black-hided cattle and carcass specifications."

In January 1978, Colvin began working with the federal government and packers simultaneously, trying to create a system for the first program of branded fresh beef that was also USDA's first major certification program. Colvin was trying to tap into the government's beef carcass data service, but the lead inspector was focused on doing his job for all producers and consumers rather than for this Angus group. Although USDA seemed to raise some obstacles at first, in February 1978, officials agreed to the proposed specifications.

With the scientific foundation established, they needed a marketing plan.

Dr. Bob VanStavern, a meat scientist from The Ohio State University, is asked to create the brand's science-based quality standards. VanStavern has long wanted to encourage quality beef, while the supply was shifting toward cattle with low marbling.

Consultants Who Filled the Gaps

As Colvin sought the expertise of VanStavern for meat standards, he now needed assistance in marketing, branding, and public relations—three areas identified that would shape the company. From the beginning, a strong logo was considered a top priority. Etling and Elliot both noted that in early letters.

The committee employed the help of Belden/Frenz/Lehman, an advertising and sales promotion agency in Cleveland still doing business today under the BFL name. Johnson had worked with the firm before and had a good relationship with them. The first recommendation made was to change the program name from Angus Beef Certification to the Certified Angus Beef program.

"Certified Angus Beef just sounded better. There was alliteration to it," says Bernie Rowan, a former account manager for Belden/Frenz/Lehman. "The important thing really being the word 'Certified,' because there was an awful lot of meat called Angus beef. Certification was what made their program unique; that's why we suggested they put 'Certified' first."

Rowan also created the first logo, but recalls the project was exceptionally easy. As a graphic designer, Rowan was accustomed to sitting in long meetings about what cli-

Exclusive Report on the details and workings of the New ABC Program by the Illinois Angus News

New, Aggressive Angus Merchandising Program Launched

A group of dedicated Angus breeders and business men, in cooperation with the American Angus Association have just launched a program that has the potential to contribute more to "Angus Merchandising" than any program ever implemented by a beef registry association.

The beliefs of these people, confirmed by extensive interviews with packers and retailers, are that a certain segment of consumers are willing to pay a premium for top quality beef — top quality Angus beef.

According to the executive director of the new program, Mick Colvin, former Regional Manager for the American Angus Association, "by 1980 approximately 60% of all beef consumed will be in the form of hamburger. The vast majority of the other beef will be in the form of steak.

It is our responsibility, as producers of the highest quality beef, to afford consumers the opportunity to purchase this quality product. With the present trend over the last 20 years of lowering quality standards for beef, by the year 2000 consumers will have forgotten how flavorful, tender and delicious beef can be.

What will be even worse for Angus breeders is that the consumers who desire high quality beef will not know where to buy it. Colvin feels that the timing of the program is perfect. He said his group has seen more deterioration and **variation** in beef quality during the past 12 to 18 months than at any previous time.

This variation can be expected when we consider the vast majority of all types of fed beef — both beef and dairy types — that qualify for the present U.S.D.A. choice grade.

As Angus breeders we have a challenge Colvin emphasized, we are producing the "Rolls Royce" of the entire beef industry, but when our product is marketed it is tossed into the "lot" with all other "brands," and Angus producers and feeders must accept the "average" market price under today's marketing system.

This program will simply maintain the quality "Angus" identity after the hide is off for those steers whose carcasses meet the established quality standards as authorized by the program. These carcasses are then certified and rolled, before they are moved to the high quality retail outlets — food stores — and high quality restaurants.

This will afford the consumer that recognizes and desires quality beef the opportunity to purchase it with assurance of consistency and dependability of quality. Colvin realized that this is not an easy task, and that he has a monstrous challenge on his hands in getting this type program under way.

Here is how the program will work. The program is called "Angus Beef Certification" — ABC.

Producers (commercial cow and calf men) who use registered Angus bulls recorded in their names and on a partial Angus base cow herd will "certify" that these designated and identified calves qualify as ABC calves. This certificate is passed on to the feeder who in turn will "certify" that the calves were fed a ration that meets minimum requirements set forth by A.B.C.

The packer will maintain identity on the cattle and the carcasses that meet the quality standards established for the program, will then qualify for the A.B.C. roll, and merchandised through the selected retail outlets.

The producer will use exclusive numbered tags to identify all qualifying calves. With a computerized listing of all tags, availability, and consistency of supply will be consistently monitored.

This program has been labeled Phase II, and will require approximately three years until the first load of A.B.C. can be delivered to the retail outlets.

To take advantage of the 3-year time lag the group will initiate an interim program labeled "Phase I." This Phase I program will be a pilot project scheduled for Northeast Ohio. This program will afford the A.B.C. group the opportunity of working with the two final steps of the program in solidifying procedures from packers through the retail outlets.

Phase I will identify predominately Angus steers in selected packing plants the carcasses that qualify under the same standards as Phase II then be rolled A.B.C. and move to the retail outlets. The quality standards that all carcasses must meet in both phases are:

1) Modest marbling (med. choice) and higher.

2) Yield grade %, 3.5 and better.

3) Established criteria for color and texture.

Colvin is quick to recognize the problems associated with

If the new ABC Program works, packers will be able to identify the Angus carcasses from other breeds right in the cooler!

MICK COLVIN

PAGE 12—ILLINOIS ANGUS NEWS

APRIL, 1978

When Certified Angus Beef debuts, American Angus Association public relations director Keith Evans sends press releases to agricultural, food, and meat trade journals. The media find the program new and exciting. The results are high pick-up rates of Evans' releases.

ents wanted. Whether this group didn't have the budget for such meetings or they just didn't want to waste time, he says, they were very decisive about what they wanted. Only red and black colors would be used in a logo that must prominently feature the words "Certified" and "Angus Beef." The main graphic was to be an Angus bull, later changed to a steer, according to minutes from February 16, 1978, when the board also approved the name change to Certified Angus Beef.

1977

Mick Colvin becomes Association staff representative to committee

Board approves program in November

Committee names Colvin as executive director in December

The first logo is displayed at Renzetti's IGA in Columbus, Ohio. Many store owners do not want to take a risk, but Bob Renzetti understands and appreciates Certified Angus Beef. He has long searched for consistently high-quality beef, and he sees marketing value in being the first store to carry the brand.

"I worked up a few designs, and about two or three weeks later, we went back for another meeting and had three versions. One was accepted. And off and running we went," Rowan says. "It was a very quick decision, because the answers were obvious. Fortunately, we came up with something they liked."

Belden/Frenz/Lehman also helped with marketing materials. In March 1978, the board approved 5,000 tent cards, banners, counter cards, and eight-page booklets for $5,606.02. These materials were sent to possible licensees and to the press. Through the efforts of the Association's Evans and Susan Scott, a part-time communications coordinator for the Ohio Poultry Association who accepted a press officer role for the brand, the program generated interest from *Farm Journal*, *Successful Farming*, *BEEF*, *Columbus Citizens Journal*, and *The Columbus Dispatch*.

"It seemed like every publication in the nation had a story about it getting started," says Bob Neumeyer, who served as chairman of the promotion and advertising committee for the Association. "I estimate we were getting at least $100,000 worth of promotion."

After USDA grader acceptance, the first carcass is rolled with the brand's logo at Val Decker Packing. The beef is sent to Renzetti's IGA.

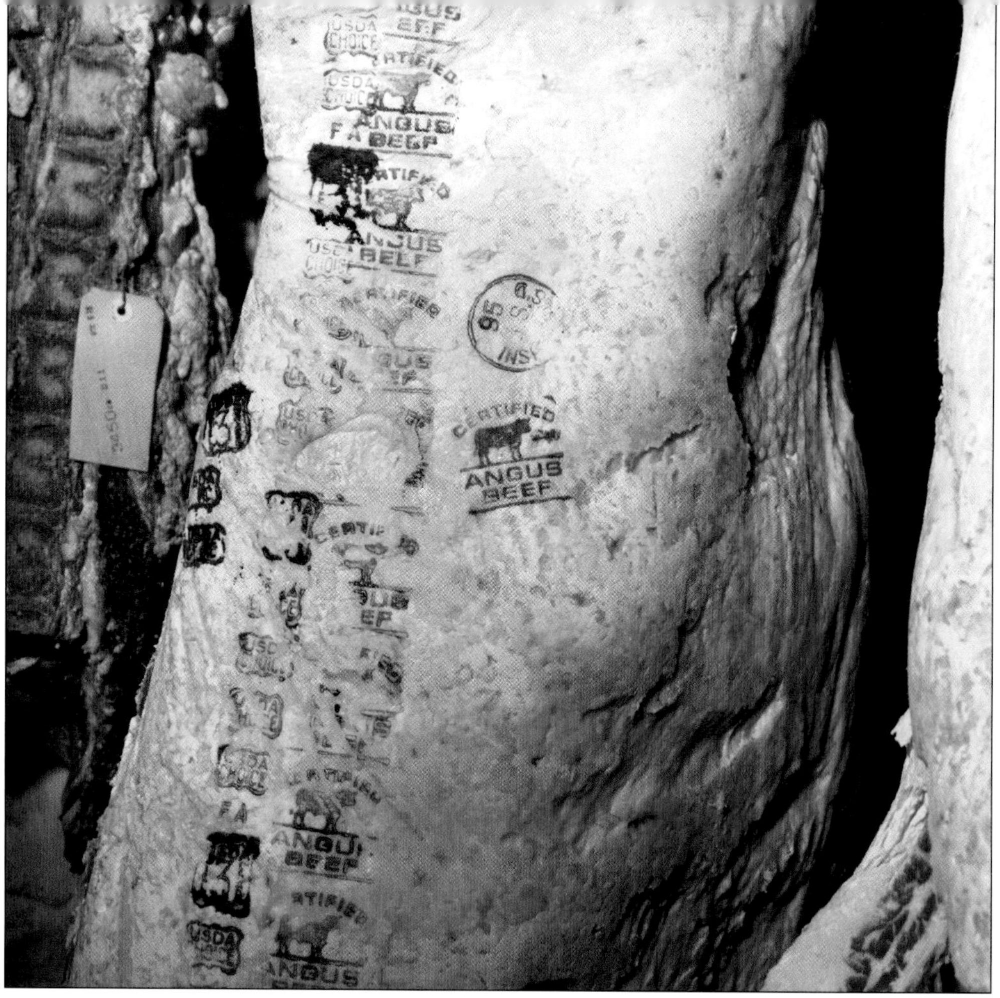

The First Licensees

Throughout 1978, the brand struggled to land its first licensee. During the first five weeks of searching, Colvin was told "no" by numerous companies. One corporate office expressed concern about paying three-cent premiums in light of the prevailing beef market that spring. Perhaps most disheartening was thinking you had a chance and suddenly having the door slammed in your face. That happened with a three-store outfit in Ohio.

The owner was a big believer in Angus cattle and claimed to understand the breed's capability of producing superior quality meat. Shown that spark of interest, Colvin determined that the store's meat packer could pull enough qualified Angus cattle to satisfy projected demand. The small packer from Perrysville, Ohio, had planned to go with Colvin to a USDA "get started" meeting when the store owner changed his mind. He feared the Angus pilot program would not have the permanence for a retailer to consider. It was a major blow to Colvin's spirits.

"After five weeks of a conscientious effort to locate a retailer for the brand, it is very discouraging not to be able to report that we have them on the dotted line," Colvin wrote to committee members in June 1978. "I realize that three cents per pound, to the committee, seems like a token amount compared to the value that the program provides to retailers and consumers. I feel confident that consumers will be willing to pay a premium for the consistency of quality beef."

At 8:35 p.m. on October 18, 1978, Ella Whitt purchases the first Certified Angus Beef product. To commemorate the event at Renzetti's IGA in Columbus, Ohio, Governor James A. Rhodes issues an official proclamation launching the country's first fresh beef brand. Attending the launch are Mr. & Mrs. Bob Renzetti, Sr.; Gilman Stewart, Association president; Bob Renzetti; Rhodes; Whitt; C.K. Allen, Association executive secretary; and Pat Lynch, checkout clerk.

When writing to retailers who were considering the program, Colvin talked about what high standards it would follow. The specifications that made this program unique were not always easy to grasp. In a letter to Bill Valentine, of Parkway Quality Meats in Berea, Ohio, Colvin tried to clarify the simplicity and benefits.

"You would continue purchasing your beef from the packer on the same pricing scale as in the past," Colvin wrote. The Certified Angus Beef brand "is evaluated and selected only after qualifying under higher quality standards than those used in the best selecting programs, thus assuring your customers the consistency of high-quality beef."

The brand started by focusing on small stores and packers in Ohio. A common response to Colvin's pitch was that nobody has complained about the store's beef before. To which Colvin frequently responded: "History has proven over the years that when a large chain store can offer the same product as a Mom & Pop or specialty store, the larger chain will usually win the battle because of higher sales volume and lower prices. However, if a specialty store in the right area can provide a higher quality product and promote the quality rather than price, their success can be virtually unlimited. A merchandising program based on quality and certified for that quality level is an assurance of customer satisfaction that is extremely difficult for larger chain markets."

Colvin's luck changed after VanStavern recommend Bob Renzetti of Renzetti's IGA in Columbus, Ohio. In Renzetti, Colvin found a smart businessman who had been

Mick Colvin (right) and Bill Decker inspect a carcass graded for the Certified Angus Beef brand. His family's Val Decker Packing became the first licensee. Production begins October 6, 1978.

Inside Mick Colvin's home office, the official Certified Angus Beef headquarters, Ed Elliott (left) and Fred Johnson show off the new Certified Angus Beef logo.

The Logo Creator

When another corporate logo designer asked Bernie Rowan what was the best accomplishment in his forty-plus-year advertising career, it was an easy answer: creating the first Certified Angus Beef logo.

"The design itself has so many good features about it, and the program was so exciting and different. It just made it one of the best things I ever did," Rowan says. "Knowing the people behind this, like Fred Johnson and Mick Colvin, I knew that they were going to do it right and really make it successful."

The logo has changed over the years, but the core elements are still there. When Rowan encounters it in the supermarket or in a restaurant, "I get a kick out of still seeing the logo—even though it has been developed now into a multicolored logo, which is very attractive."

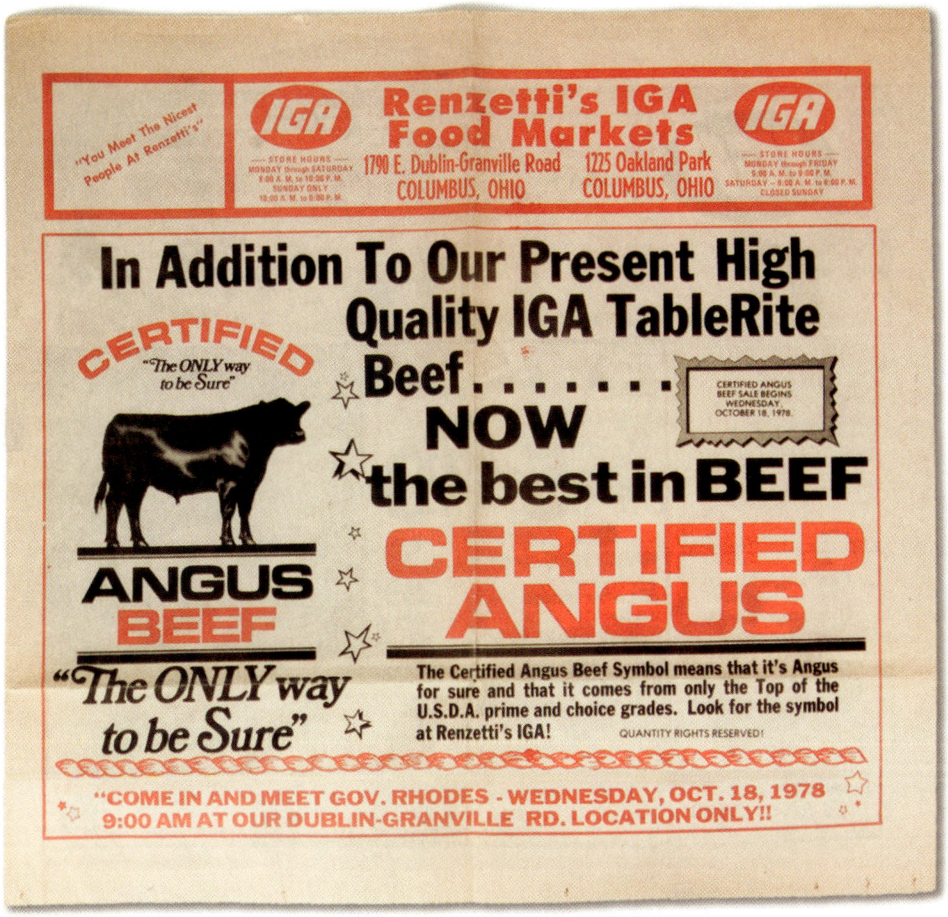

The first circular invites shoppers in Columbus, Ohio, to Renzetti's IGA.

searching for consistently high-quality beef in his meat case for years. He understood the marketing impact the Certified Angus Beef brand could have on his stores. Colvin confirmed Val Decker Packing Company, Piqua, Ohio, could access enough qualifying Angus cattle to supply Renzetti's stores.

When Renzetti and Bill Decker became the first two licensees, they proudly donned the Certified Angus Beef logo.

"They were pioneers," Colvin says. "They were really taking a chance because nobody had proved this would work."

On October 6, 1978, Colvin and VanStavern went to the Val Decker packinghouse to assist in the processing and certification of beef from the first load of cattle for the program. Evans, who went along to take pictures, recalls focusing his lens on the graders as they stamped the "A" on the first carcasses. He sent the photos and a press release to about fifty newspapers and journals.

Evans remembers promoting the event was relatively easy. "For the first store opening, the press materials went to all the livestock publications, packer publications, restaurant publications, and local media," he says. "When we actually sold the first product, everybody used the picture and the story."

That special day was October 18, 1978, at Renzetti's IGA on Dublin-Granville Road. Ohio Governor James A. Rhodes issued a proclamation officially launching the Certified Angus Beef program in the United States. It was a momentous occasion. On hand

Other Programs

After national press hit the mailboxes, two people wrote the American Angus Association describing their programs, which they considered similar. Both offered their assistance.

George Wilson, of Hygrade Food Products Corporation in Detroit, Michigan, wrote of the Red Triangle Beef program attempted in the 1940s and early 1950s. It was a cooperative between Angus breeders and Canadian packers in Toronto.

David Hewlett, of the Beef Direct Marketing Project from South Dakota, was trying to sell beef straight to the consumer. Funded through the Farmer-to-Consumer Direct Marketing Act of 1976, Hewlett's program focused on high-yielding cattle. In a letter to Colvin, Hewlett says they were trying to "reward breeders of efficient cattle," but he made no mention of beef taste.

In addition to widespread print media coverage of the first pound sold, Columbus' Channel 10 covers the event at Renzetti's. Mick Colvin (top photo) speaks with a reporter in front of the meat case. The reporter also talks with a consumer about the product.

were Johnson, VanStavern, Colvin, Cannon, Evans, and C.K. Allen, the Association's executive secretary.

Evans captured the ribbon cutting in a flawless black-and-white image of the Ohio governor handing first customer Ella Whitt her package. In the photo, everybody is smiling as if they know they are a part of something special. Afterwards, Colvin treated the guests and press to a fine lunch at the Wine Cellar restaurant across the street, where they served—of course—Certified Angus Beef steaks.

Colvin went home that night to his century-old farmstead and the official company office in West Salem, Ohio, thrilled with the accomplishments. After all his struggles to reach distributors, retailers, and restaurants, the brand had finally sold its first pound. Things were looking up.

But as people were learning about the Certified Angus Beef program for the first time and wanting to put the premium product in their meat cases and on their menus, it seemed that the worst possible thing happened. The USDA terminated the program.

1978

Colvin begins "Phase I" product development; Dr. Bob VanStavern hired as consultant

"Certified Angus Beef" name and logo approved by newly renamed Certified Angus Beef Committee

U.S. Department of Agriculture (USDA) sends letter of approval

Production begins at Val Decker Packing Company

Ella Whitt makes first purchase at Renzetti's IGA

USDA cancels program citing unclear procedures

USDA certification adds ultimate value to Certified Angus Beef by identifying carcasses that meet all specifications for the brand. USDA graders act as an independent third party to give the brand utmost credibility.

2 A Fight to Survive

Three days after the Certified Angus Beef brand made its first sale, *The National Provisioner* ran a story about the program. The meat industry's authoritative publication used the logo, quoted Association chief C.K. Allen, and outlined how the program worked. The magazine, still in business today, essentially ran Evans' news release verbatim. It was the kind of media placement that public relations personnel dreamed of because the article touched many decision makers, including Charlie Robinson, a twenty-nine-year-old executive at distributor Oxford Trading Company in the Boston area.

"I saw that article and wanted to get involved. I knew this program would distinguish us from all our competition by providing our customers with consistent quality," Robinson remembers. "I went to my boss Jack Saltman and said, 'Look at this article. This Angus program is going to be the future.'"

Indeed, that story and other national media coverage seemed to prove the Association had achieved the improbable and was on its way to changing the landscape of the meat industry. About the same time Robinson learned of the brand's lure, benefits, and potential, some USDA officials learned about the program for the first time and shut it down. There was never a consistent, official explanation from USDA, but a December 1978 *Food Chemical News* article noted, "it was approved by inspectors in the field, but never by Washington headquarters, where all labels must be approved. ...The Department has shied away from breed labeling in the past because of difficult control problems, which may require registered papers proving that the carcass was of a certain breed."

Association executive secretary C.K. Allen (third from left) supports Certified Angus Beef from its first days and fights like a bulldog to keep it alive when the USDA temporarily cancels it. Allen is pictured with Mr. and Mrs. Bob Renzetti Sr. and their son, Bob, after the first sale at Renzetti's IGA.

However, Colvin says no such communication was ever given to him, Allen, or any representative from the Association. Some blame Carol Tucker-Foreman, the assistant USDA secretary during President Jimmy Carter's administration and known consumer advocate, for single-handedly canceling the program. It was believed that she saw it as a consumer fraud. Thirty years later, she denies that, saying the program's 1978 demise was a combination of pressure from other cattle breeders, bad timing, and inspector manpower.

Calling the 1970s, "the cattle wars," Tucker-Foreman says, "The cattleman's worst enemy was the other cattleman down the road. If you wanted to do poultry inspection all you had to do was get the five biggest companies in the room and they made all the decisions. If you wanted to get red meat regulation, you had to go through agonies of the small producers against the big producers and the small packers against the big packers." She says that non-Angus breeders argued against the virtues of the Certified Angus Beef program because it would have hurt the sale of their cattle.

"The people lobbying against it were not consumer people. They were the cattlemen who weren't going to get [their beef] certified," Tucker-Foreman says.

That story rings true for Mike May, a retired Branch Chief for USDA Standardization. May says the Continental cattle breeders knew the Angus influence improved meat quality. He claims they worried about segregating Angus from their breeds, because "that was going to shove exotic breeds off in the corner. If you separated Angus, they knew it would make them look bad by comparison."

USDA inspectors focus on beef safety and plant cleanliness, while graders segregate beef into grades based on marbling, maturity, and yield. Both are absolute necessities for Certified Angus Beef, arming the brand with independent meat quality assurance and objectivity.

USDA Structure

The Certified Angus Beef brand works with USDA's Food Safety and Inspection Service (FSIS) and its Agricultural Marketing Service (AMS). Both divisions operate separately and have their own objectives.

FSIS, commonly referred to as "Inspection," ensures that the nation's commercial supply of meat, poultry, and egg products is safe, wholesome, and correctly labeled and packaged. Its services are mandatory.

On the other hand, AMS offers quality grade standards, grading, certification, auditing, inspection, and laboratory analysis.

"We've always viewed our role as a service to the industry. We're a voluntary service. No one has to use our service," says Larry Meadows, chief of the Meat Grading & Certification Branch of AMS. "We're helping farmers market their products."

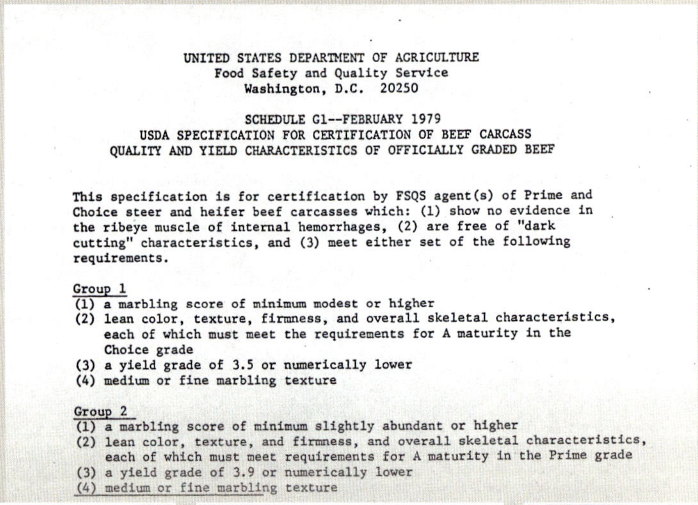

UNITED STATES DEPARTMENT OF AGRICULTURE
Food Safety and Quality Service
Washington, D.C. 20250

SCHEDULE G1--FEBRUARY 1979
USDA SPECIFICATION FOR CERTIFICATION OF BEEF CARCASS
QUALITY AND YIELD CHARACTERISTICS OF OFFICIALLY GRADED BEEF

This specification is for certification by FSQS agent(s) of Prime and Choice steer and heifer beef carcasses which: (1) show no evidence in the ribeye muscle of internal hemorrhages, (2) are free of "dark cutting" characteristics, and (3) meet either set of the following requirements.

Group 1
(1) a marbling score of minimum modest or higher
(2) lean color, texture, firmness, and overall skeletal characteristics, each of which must meet the requirements for A maturity in the Choice grade
(3) a yield grade of 3.5 or numerically lower
(4) medium or fine marbling texture

Group 2
(1) a marbling score of minimum slightly abundant or higher
(2) lean color, texture, and firmness, and overall skeletal characteristics, each of which must meet requirements for A maturity in the Prime grade
(3) a yield grade of 3.9 or numerically lower
(4) medium or fine marbling texture

A

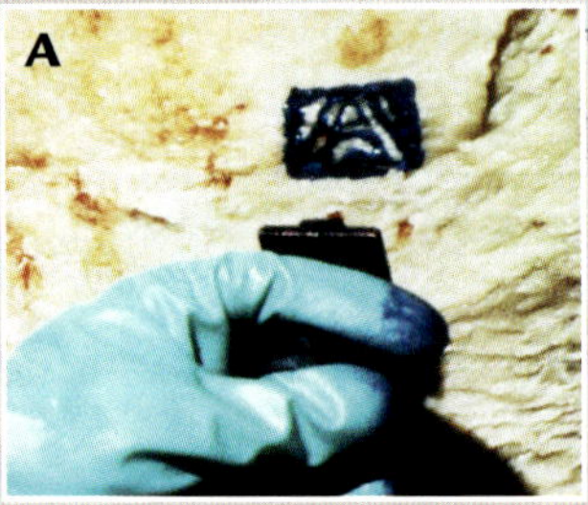

B

C

USDA Certified

Harold Etling's original concept for Certified Angus Beef requires a rigid inspection program. Etling, Colvin, and committee naturally turn to the U.S. Department of Agriculture, which has beef graders in each U.S. packinghouse. Consumers likely recognize their work in the form of USDA beef grades—Prime, Choice, and Select—on retail meat packages and some menus.

In 1978, Certified Angus Beef asks USDA to go a step further by evaluating beef for more specific marbling and maturity standards, among others. These strict standards are eventually set forth in Schedule G1, the first procedures for a "USDA-certified" brand, providing impartial evaluation by a third party. Certified Angus Beef is the only Angus brand both certified by USDA and owned by the American Angus Association.

Photo A: At the packinghouse, trained plant employees separate predominately black-hided steers and heifers from other cattle. Black hair coat is a predominant trait of the Angus breed. Their carcasses are stamped with a large "A" for further evaluation.

Photo B: Trained employees mark "G1" on carcasses to be considered for the Certified Angus Beef brand. USDA graders evaluate each of these and certify those that qualify. Only USDA graders can apply "Accepted as Specified" overtop the "G1." This product is marketed as Certified Angus Beef.

Photo C: In the program's earliest days, the logo is rolled onto each approved carcass. Modern grading procedures allow for "Accepted as Specified" to be the official USDA certification.

When the program is launched, media coverage captures the interest of executives at Oxford Trading Company. (Left to right) Charlie Robinson, operations manager, and Jack Saltman, president, embrace the opportunity to become the first distributor. Norm Slawsby is among the first sales associates to offer the product to customers.

In addition, the program fell into bad timing. In 1975, meat packers in Los Angeles County pleaded guilty to bribing graders. That crime was on taxpayers' minds after *60 Minutes* exposed it. "They were bribing the graders out on the West Coast because the meat had a hard time meeting the Choice grade," Tucker-Foreman remembers. "The Angus program was birthed in the middle of a period of a lot of controversy in the packing industry."

She says a special committee oversaw the development of the program and she made her decisions based on her team's advice and opinions. In correspondence with company representatives, Tucker-Foreman wrote: "We are concerned that these kinds of certification programs will rapidly expand and use up already limited inspection resources." After that letter to Johnson, Colvin submitted data to the USDA branch showing it would only take inspectors three to four seconds to sort off A-stamped carcasses, but that did not seem to help. USDA Branch Chief of Standardization Herb Abraham, who granted the Certified Angus Beef brand's approval in 1978, says inspectors just didn't want the extra work.

"The objections had to do with the additional work on the kill floor at the time of slaughter," Abraham says. "When the program stopped, it was because inspectors didn't want to do it."

May recalls it this way: "The inspectors at that point, to my knowledge, were not overloaded with what they had to do. It was a lame excuse."

Tucker-Foreman contends inspectors are law-enforcement officials who ensure meat products are safe, not adulterated, and accurately labeled. "The labeling responsibility for

Diggs Packing is vital to Certified Angus Beef's retail success in the Show-Me State. Brand leaders (left to right) Dr. Bob VanStavern, Fred Johnson, and Mick Colvin visit the plant soon after it licenses in 1979.

1979

After much consultation, USDA approves "G1" evaluation procedures

First retailer after USDA re-approval: Eastgate IGA Foodliner; Diggs Packing supplies product

First distributor: Oxford Trading Company

First grocery chain after re-launch: Hedges & Hafer Markets

FSIS [USDA's Food Safety and Inspection Service] involves nutrition and safety, not meat quality or marketing claims," she says. "It is not the role of meat inspectors or the meat inspection staff to assist in advancing a marketing program for a particular product."

That was where the waters became muddied. In one interpretation, there was nothing wrong with having inspectors sort the cattle. After all, a 1977 Meat and Poultry Inspection Bulletin 77-12 stated: "MPI approves acceptable plant quality control programs that can be effectively monitored by inspectors. These programs may cover ... percentage labeling" among other things, such as percent pickup of enzyme solution and yield. In addition, under USDA's Beef Carcass Evaluation Service, inspectors were charged with supervising the evaluation of live animals. Add that to the fact that the Certified Angus Beef brand had already been approved and proven to work, and you have a group of Angus producers who just did not know why their program had been canceled.

"Our plan is to maintain the identity of Angus cattle through slaughter and then select only those that have young, high quality, lean carcasses to carry our Certified Angus Beef label. USDA has existing programs that they offer commercially to the beef industry such as the USDA Carcass Data Tag Service, the Acceptance Program, and the Beef Carcass Evaluation Service," Allen wrote to Secretary of Agriculture Bob Bergland on December 7, 1978. "We have had outstanding cooperation with the Standardization, Marketing, and Grading branches of USDA, but I have encountered one obstacle after another with the Meat and Poultry Inspection Division. All that we have requested Meat and Poultry Inspection officials to do is to fulfill their obligation to supervise the transfer of identity in USDA's Beef Carcass Evaluation Service and to review our request for label approval objectively, based on the merits of our program. Several USDA

Today's USDA Relationship

Although the relationship started on rocky terms, today, the USDA and Certified Angus Beef partnership is a mutually beneficial one. In fact, USDA's Mike May presented the company with the 1995 Livestock and Seed Division Industry Recognition Award, recognizing its pioneering efforts and contributions in developing quality marketing systems for beef.

USDA has provided independent, third-party approval for the brand's quality from day one. As the brand celebrates its thirtieth anniversary in 2008, it honors the USDA for its long-standing commitment to branding fresh beef. President John Stika (right) presents a branding iron to Larry Meadows, chief of Meat Grading & Certification.

May allows there were some hiccups here and there, but very few major incidents occurred during his tenure with that department, which ended in 1996. "The issues ranged from plant to plant, but it was usually a communications issue at plants," he says. "Once the program got put into play, it was smooth."

Company president John Stika says even in the early days, there was a great relationship with the grading and certification branches of USDA.

"One of the things you notice when you go into a plant is the USDA graders take a sense of ownership in Certified Angus Beef," Stika says. "They will not allow anything that doesn't meet the specs to get into this program, and that's what we want."

Today, the two organizations work closely together with licensed packing partners, making sure the end product meets the specifications. The big difference now is that USDA certifies many other programs today, more than eighty beef programs in 2009.

Meadows says he believes the sixty-plus other Angus programs would not exist had not Certified Angus Beef jumped the major hurdles in the early days. Staff within the brand share mutual esteem for their USDA colleagues.

"USDA has added to our credibility as a third party that's known for quality," Stika says. "I don't believe we would find ourselves in the same position we are in today if the USDA wasn't involved. They have done an outstanding job of applying those specifications across multiple packers."

As for the inspection side of USDA, Stika believes their role continues to assure the American public that Certified Angus Beef is the real deal. Inspectors oversee the process of the meat going from the carcass to the box to the package. Stika says he's thankful for such monitoring practices. "We don't want a low-quality product with our logo. We're a program based on fact—not myth and marketing," he says. "We tend to find ourselves in a strong position with USDA inspection and labeling, probably as a result of the hard work put forth by Mick and others early on."

Above left: With final USDA approval in April 1979, Certified Angus Beef enjoys immediate growth in Missouri. Eastgate IGA Foodliner sells its first package to Liz Menzel.

Above right top: Angus cattle are displayed outside the store to encourage product sales.

Above right bottom: And on local television, Mick Colvin and storeowner Dale Cornelison discuss the program's unique and added value for consumers.

employees along the chain of command between Washington, D.C., and the packing plant inspection level do not know their responsibilities."

In one of his many well-written letters to government officials, Allen thoroughly described the need for quality beef and this program's scientific standards. The product was a hit with consumers and grocers alike. "It attracted national coverage within the food industry, *CBS News*, daily newspapers, and agricultural press," Allen wrote. "Some of the personnel within the Meat and Poultry Inspection Division apparently became alarmed. They did not understand the program, had not heard about it, yet they did not call us to get information. Without even the courtesy of a phone call, they instructed the plant inspector to confiscate our carcass roll on November 1, 1978, at the Val Decker Packing plant."

A week after the cancellation, Colvin, Allen, Dr. Jerry Lipsey, a meat scientist affiliated with the Association, and attorney Lee Badger met with several USDA officials. Company representatives detailed the program, explaining the checks and balances that made it work for packers, graders, and inspectors, as well as the need to meet the consumer demand for quality. They also offered to modify the procedures to meet whatever USDA's concerns were. They even completed an MP Form 480, now held up as a requirement but not mentioned prior to the cancellation.

USDA officials swiftly came to the point, telling the Association representatives that the program was too detailed and had too many checks and balances to overturn

Thurs., Oct. 30, 1980 The Drovers Journal Page 19A

Certified Angus Beef sells program to first restaurant

ST. JOSEPH, Mo. — The Maple Root Inn at Coventry, R.I. became the first restaurant in the United States to sell Certified Angus Beef (CAB) recently, reports C.K. Allen, executive vice president of the American Angus Association.

The CAB program, started in Oct. 1978, is the first time that members of a beef breed organization have merchandised their breed's beef directly to consumers through supermarkets and restaurants.

CAB is "an outgrowth of the decline of the USDA beef quality grading standards, combined with the desire of many Angus breeders for a program to merchandise and identify superior Angus beef carcasses," says the association.

Angus cattle are identified and certified at the packing plant. Breed identity is maintained from the live animal to the individual retail cut. Carcasses must pass these seven specifications:

- Have at least modest (average Choice) marbling.
- Fall within the "A" maturity range.
- Yield grade 3 or better for Prime and no higher than the midpoint for the 3 grade for Choice.
- Have lean within the "fine" texture range.
- Have fine to medium marbling texture.
- Color of lean must be slightly dark red or lighter.
- Meat must be moderately firm with no indication of softness.

Participating packers and retailers signed an agreement with the American Angus Association defining the program. The packer obtains the Angus cattle and sells the beef directly to the retailer. Both are subject to unannounced inspection.

The Angus beef is supplied to the Maple Root Inn and other supermarkets in the Boston area by the Oxford Trading Company of Boston.

New vaccine is praised by officials

RICHMOND, Va. — The U. S. Animal Health Association, (USAHA), has applauded a new Brucellosis vaccine recently made available.

"We have now entered into a cooperative spirit between the industry, the USAHA and the USDA," said Bert Hawkins, an Oregon rancher and president of the USAHA.

The new "reduced dosage Strain 19" is expected to control a disease that USAHA estimates causes a $40 million loss each year.

Also known as Bang's disease, Brucellosis causes abortion, weak calves, reduced milk production and impaired breeding. It also causes undulant fever in humans.

Most states have less than .5 percent of their cattle infected with the disease. However, Oklahoma, Texas, Mississippi, Louisiana and Florida have 1 to 4 percent infection.

The USAHA is a non-profit organization. It works with health officials, veterinarians, producers, livestock organizations and some foreign countries in controlling livestock diseases, and is an advisory group to the USDA.

In June 1979, Oxford Trading licenses the first restaurant, the Maple Root Inn in Coventry, Rhode Island. The classic diner receives press coverage and regularly advertises the brand in its local paper.

the cancellation. Weeks later, Allen would receive additional excuses as to why the program was cancelled and why it could not be reinstated. One group said the roll used to stamp the Certified Angus Beef logo on the carcass was not approved, and skeptics in another division said they did not know about their responsibility to supervise transfer of brand identity.

Meanwhile, the Association never stopped fighting. Board president J.E. Horton Jr. requested the assistance of Congressmen Bill Nichols and Ronnie Flippo (Alabama Democrats), and Senator Herman Talmadge (Georgia Democrat) to put political pressure on USDA to approve the program.

"The Food Safety and Inspection Service of USDA has raised objections to our project—objections which are not specific and which we do not understand," Horton wrote to the Alabama Congressmen who, along with Talmadge, were considered friends to farmers. "Because of our inability to communicate with the upper echelon officials, we are asking you to contact these people on our behalf to see if they will meet with us to help resolve their objections. In our opinion, we have a well-designed, closely monitored program that ultimately can be of great benefit to consumers of beef and to producers of Angus."

The strong writing trio of Johnson, Allen, and Colvin wrote Tucker-Foreman and Deputy Assistant Secretary of Agriculture Sydney Butler on what seemed like a weekly basis requesting an in-person meeting. These letters to USDA officials showed their frustration with the bureaucracy, too. Johnson wrote to Tucker-Foreman: "I find it difficult to believe that you, an outspoken consumer protection advocate, can allow the USDA to needlessly delay approval of the Certified Angus Beef program. Obviously, you are not fully informed about the program or it has been misrepresented to you." In a

What Does Tucker-Foreman Think of the Brand Now?

Today, Carol Tucker-Foreman is a big fan of the Certified Angus Beef brand. She calls the Association members who pushed for its acceptance, "pioneers."

"The Angus people came up against problems similar to those of the organic growers twenty-five years later," says Tucker-Foreman, who now serves as a distinguished fellow for the Food Policy Institute and is still considered a renowned food-safety consumer advocate.

She says the Certified Angus Beef program gave other organizations a blueprint for creating a certification program. But she still contends that USDA's Inspection arm should not be asked to work in marketing programs.

"As it turns out, the Angus Association has prospered without having it built into meat inspection," she says. "They came up with a good idea and sold it pretty much on their own."

similar letter to Tucker-Foreman, Allen said: "We don't want to waste your time nor do we want to get USDA involved in something that will embarrass them at a later date."

Colvin says their letters might have been a little antagonistic, but adds they also represented how strongly the people believed in the program. "C.K. was young and aggressive. Fred and C.K. could come across a little too strong in their letters, but that made them good for us," he says.

When 1979 came around, those letters still hadn't opened any new doors for Certified Angus Beef. The group discussed what seemed to be coming: the inevitable end. But Etling, Elliot, Johnson, and Frank had all weathered their share of major business deals. Young and full of energy, Allen and Colvin were too stubborn to give up. They concluded more than once that there was still some fight left—there was too much at stake just to give up.

"I was younger and more foolish then. That's probably why I didn't throw in the towel," Allen says. "We were going up against a major government entity, trying to change an Agriculture Department directive after it had been issued to cancel the program, but we also had a lot of help in the industry. Politicians sat down with Carol Tucker-Foreman at her breakfast table and explained the program to her. We were getting lots of help that probably was never really recorded and we also had other people contacting their congressmen and their senators, trying to come in through any person that knew anybody in the White House."

That strategy seemed to have paid off. On January 11, 1979, USDA Inspector John McKelvey granted the Certified Angus Beef program a six-month temporary status for use at Val Decker Packing Company. McKelvey later assured Colvin there would be no problems with expanding to other plants as long as the program followed procedures outlined in MPI Bulletin 77-12. But that glimmer of hope was shattered when USDA refused to initiate its own Acceptance Program, forcing Colvin to cancel plans with retailers in Ohio and Missouri. Once again, the Certified Angus Beef program had gained momentum. And once again, USDA canceled it after granting approval.

"You wanted to cry. The USDA was dragging their feet for political reasons and had no basis in fact, so it left you feeling helpless," VanStavern says.

However, once again, a setback would not slow down the group's resolve. By February 1979, Colvin, who practically lived in Washington, D.C., at the time, worked with officials to establish procedural requirements for USDA Certification as set forth in Schedule G1. It was to be the Association's responsibility to train licensed packing plant employees to identify black-hided cattle. The licensee, through an evolving system of designation, would provide meat graders with the specific cattle and carcasses eligible for the program. At first, that required a written and dated statement of certification with lot and number of head in each pen, including a copy sent to Inspection. As for carcass evaluation, plant employees would identify each eligible animal by removing one ear, placing it in a plastic bag, and attaching it to the carcass.

Obviously not well suited to significant volume, the original process was later changed with USDA approval to streamline live and carcass identification by using an "A" stamp, for "Angus," as hides are removed. Then as now, inspectors observed the

process during the course of their normal duties, and graders later stamped an accepted mark on each identified carcass qualifying for Prime or premium Choice.

On March 8, 1979, Tucker-Foreman approved the program under the supervision of Grading, and Eddie Kimbrell, the deputy administrator of Commodity Services, accepted that responsibility. But Kimbrell did not support the first draft of the procedures. Colvin added the opportunity for graders to observe the cattle before harvest. On April 2, 1979—exactly six months after the cancellation—Kimbrell gave the Certified Angus Beef program its final approval. This time, there would be no sudden turn of events that forced a cancellation.

After the acceptance, Colvin got back to the job for which he was hired—to license retailers, packers, and restaurateurs. And a young New Englander who just entered the meat business found the niche that would make his career. Charlie Robinson and Oxford Trading Company became the first licensed distributor in June 1979. Robinson knew right away, "I wanted to change the whole direction of Oxford Trading Company; I'd like to just sell Certified Angus Beef product, and it all started from that little article in the *Provisioner*."

Certified Angus Beef gives meat managers an alternative to commodity USDA beef. The company provides in-store marketing materials, giving consumers of all ages a recognizable mark of quality.

Fred Johnson is fondly remembered for his genuine personality and his passion for the brand. While Mick Colvin serves on the front lines pitching retailers and distributors, Johnson makes phone calls to Angus producers who vote on the program's future and writes poignant letters to government officials.

Johnson's Letters

Sam Johnson, Fred's son, remembers the USDA and Association trials quite well. After all, he, his brother, sisters, and mother served as unofficial editors.

When the elder Johnson drafted a letter, he read it to them "unless we could find some way to escape," Sam recalls. "But I was always the slow one, so I always had to listen."

Johnson's idea was to write letters that even his thirteen-year-old boy could understand. If Sam asked a question, the subject probably needed elaboration. Mother, however, did her best to soften the chairman.

"My mother would say: 'Fred, you're being too nasty. Be polite. You catch more flies with honey than vinegar.' My mother really did tone my father down quite a bit and probably made him a lot more effective as far as winning friends and influencing people."

In 1979, Hedges & Hafer begins selling product in three stores in greater St. Louis. The region remains a top market for the brand, including licensed retailer Schnuck Markets.

Retail Boom

Soon after the USDA green light, Renzetti's IGA resumed selling the brand and Eastgate IGA Foodliner in Columbia, Missouri, became the second retail licensee. Supplied by Diggs Packing of Columbia, Eastgate sold its first pound to local shopper Liz Menzel on April 18, 1979.

Selling Angus beef directly to consumers seemed like a grand idea to packer Dale Diggs, especially since Missouri was one of the largest Angus-beef producing states. "Finished cattle are a little hard to come by these days, but a large percentage of those available in this area are Angus," Diggs said in a June 1979 *Angus Journal* article.

The excitement carried over to other Missouri towns. On July 18, Fricks Market in Union, and Central Market in Sullivan started selling Certified Angus Beef products. Despite its slow start, the brand had sold 29,538 pounds through June. By December, Hedges & Hafer opened three stores in St. Louis and St. Charles. It was steadily growing and consumers started remembering the smaller stores for their quality beef. As Colvin predicted to store owners, they were differentiating themselves in the marketplace.

The big boom came in early 1980 when John Morrell became the first licensed packer with multiple plants, creating more supply. In March, two Del Monte stores started selling Certified Angus Beef products in St. Louis. Lou Stecher Markets licensed three stores in March and in June, Southtown BiRight signed up. All the while, Robinson was working on New England retailers, preaching the scientific standards used to select beef for the brand and detailing how grocers could really boost sales with this new product.

1980

First packer with multiple locations: John Morrell & Company

Lorenz International delivers first full load of boxed beef to Oxford Trading for eight New England retailers

First retailer on East Coast: B&B Market

First million pounds sold to date

First restaurant: Maple Root Inn licensed by Oxford Trading

Besides Certified Angus Beef, the American Angus Association (pictured) tests the waters with other new ventures in the late 1970s and early '80s. Sire evaluation, the National Junior Angus Association, and Angus Journal *have also proven vital to the Association's long-term success.*

"This program gave me something that was more than making money," Robinson says. "It gave me a mission. I knew that we were selling the best beef because of the genetics and interior marbling."

But Robinson met a lot of challenges. For one, boxed beef was relatively new and people who wanted a quality product didn't trust it. People had a hard time comprehending the fact that the Certified Angus Beef product came from the top two marbling levels of Choice and from Prime. To them, Choice was Choice, he recalls. One man said dismissively to Robinson, "'Angus-shmangus!' I said to him, 'You're making a big mistake. This program is going to change beef.'"

In 1980, Robinson licensed eight retailers and the first restaurant for the company. By October, the brand had sold one million pounds. For Colvin, the seven-figure mark meant three years of hard work had finally paid off.

"That was really a big deal," he says. "We felt like we had a lot of momentum."

Unfortunately, despite great growth trends in the retail markets, the trickle-down effect the committee members hoped for was not yet evident. Angus producers were still losing market share to the Continental breeds, the Association was losing money and registrations, and some people still didn't understand the program. In June 1981, the Association voted fourteen to one to cancel the program unless it was profitable within a year.

Association Woes

While that edict was alarming, it was not surprising. As early as August 6, 1978, the Association expressed concern about the program making money. In meeting minutes, Gilman Stewart pointed out that the concept was based on at least a two-year program and that it had only been approved for one year's funding. He was a proponent of the brand and fought to secure funding for the second year, but Stewart was not on the 1981 board, when shrinking budgets were top of mind.

In addition, the Association had taken on several other new programs over the two-year span. It started the National Junior Angus Association, purchased the *Angus Journal*, started field-data sire evaluations, and issued the first *Pathfinder Report*. It had to cut costs just to survive.

"We were taking on a lot of things," Evans recalls. "Nobody doubted the Certified Angus Beef program was a good thing for producers, but those were some tough times."

Bob Neumeyer remembers well how his fellow board members doubted the program: "They'd just say we're putting money down a rat hole. We were cutting costs at the time, and some of them didn't have the confidence that the Certified Angus Beef program would make it."

Nonetheless, Certified Angus Beef committee members were absolutely devastated by the thought of being shut down by their parent company. They had been doing everything possible to make the idea a success, becoming the first program to brand fresh beef. Slowly, they were changing the market. Colvin just knew it.

Charging forward in 1982, milestones were achieved with the licensing of Beef Specialists of Iowa, the first fabricator to inventory product and the first foodservice distributor to help get their foot in the doors of restaurants. DeBragga and Spitler of New York City would become the gateway to the world's top restaurants. But come September, the Association Board was ready to determine whether to keep the program. And going in, Colvin had no idea which way it would go.

"I remember sitting out in the hallway, thinking what if this thing gets shut down. What will I do?" Colvin says. "Fred Johnson came up to me and said, 'Don't worry, Mick.' He told me he would buy the program if the Association canceled it."

Neumeyer says the discussion to keep the program was very heated, as the members debated a number of questions: Can the program really work? How will it help producers? Is there really a demand for the beef? Why aren't we in more restaurants? Can we continue to justify spending money on the program?

The board voted eight to seven to keep the program. Neumeyer believes it came down to the brand's potential and its publicity. And Colvin and Johnson. He says the board may not have always agreed with the Ohio duo, but they were so confident, smart, and hard working that they commanded the kind of respect that wins people over.

Charles Cannon agrees.

"I think the board genuinely wanted it to work, but they were just skeptical because nothing like this had ever been done before," he says. "They questioned whether it would be successful because of tremendous hurdles they had to overcome, but because of their respect for the people, they supported it."

In later years, people involved with the program back then would joke, "Now, you can't find the seven people who voted against it." But the truth is, the Certified Angus Beef brand was one vote away from being shut down. That eight-seven vote is just part of history, "water under the bridge," says Bryce Schumann, current Association CEO.

"The reason this succeeded is because of guys like Mick. That kind of dedication is rare. I am certain a lot of people told him, C.K. and everybody, that it can't be done. They weren't about to let it fail," says Schumann, who worked in the brand's feeder-packer

1981

Association votes 14-1 to cancel program if not profitable within year

"Phase II" supply development concept approved by Board

Licensing DeBragga and Spitler gives the brand access to the elite New York City market. The distributor supplies the city's finest restaurants and top culinary talents, many of whom are now considered celebrity chefs.

relations division before joining the Association staff in 2000. "And from the Association's standpoint, it was a huge commitment of resources at a time when our organization had very little. It would be hard to imagine the pressure Mick would have been under. He had a huge job, very little resources, and a board of directors saying you've got to make progress: It's got to make money; we're not going to subsidize you forever; we've got to break even.

"There were so many things that were a huge detriment to the Angus breed at that time," he continues. "And what did we do? We launched a sire-evaluation program when nobody had one. We launched a branded beef program when nobody had one. We were making all these moves while registrations were continuing to drop. But if we never made those moves, we would have never turned the corner."

The corner had been turned. And Colvin found somebody who could not only help the program grow, but make it flourish in a way he had only dreamed of.

Virginia Colvin is the first unofficial employee. During the early years, she takes phone calls, proof-reads letters, packs boxes of promotional materials, and keeps committee meetings productive with her fine cooking.

Virginia's Brownies

Mick Colvin frequently referred to his wife, Virginia, as the first Certified Angus Beef employee. She helped him send letters, set up appointments, traveled to D.C. for meetings, and did just about anything she could to help the effort.

In addition to supporting her husband administratively and emotionally, Virginia Colvin also pleased Fred Johnson and the other board members with her Blockbuster Brownies, which she supplied for every meeting.

Made with premium chocolate, walnuts, chocolate icing, and lots of love, these delightful desserts were so decadent and chocolaty that every board member ate two or three. According to Virginia, Johnson told her, "If it weren't for those brownies, the Certified Angus Beef program wouldn't have survived."

Marc Sarrazin Sr. aligns DeBragga and Spitler with the brand, becoming the first distributor focused solely on restaurant business. Originally a French country butcher, Sarrazin champions the brand in The Big Apple and in Bermuda by delivering it to world-class chefs.

3 The Berkowitz Effect

At the 1975 Marriott talks there was a fiery Florida International University professor named Leonard "Len" Berkowitz, a meat consultant and restaurant industry icon. Berkowitz, who understood the potential of Angus from a marketing perspective and marbling capacity, was intrigued.

Fast-forward seven years and the program was struggling to land big-time packers and restaurants. It seemed ironic that the program was initiated because of a bad steak in a restaurant, but its main customers were retailers. Hoping to buck that trend, Colvin was digging for help. He found Berkowitz's name in a pile of notes from the '75 Marriott meetings and read how the Floridian knew the restaurant and meat business inside and out. Colvin then called Berkowitz, who invited the director to Miami for a meeting at the airport.

"Before Virginia and I flew down there, I asked Len, 'How do I recognize you?' He said, "I'm the ugliest SOB with the light blue sports coat.'" Although it was mainly the sport coat, "there was no question who was Len," Colvin remembers.

Berkowitz and Colvin hit it off. The professor remembered the program well and was pleased to hear of its success despite the great adversity it faced. Knowing how valuable quality beef could be to his industry colleagues, Berkowitz picked up the phone and called a foodservice distributor in Chicago. When Ellard "Butch" Pfaelzer, owner of The Bruss Company, listened to Berkowitz, he knew right away that he wanted to be a part of the program.

Since the grading change in 1976, Pfaelzer had seen a decline in the volume of Prime, and there was little consistency within Choice. It became difficult for the distributor

Above: Butch Pfaelzer, owner of The Bruss Company, is an early brand supporter and pioneer. He brings early international opportunities to Certified Angus Beef staff and pushes for growth into new territories.

Top: In October 1983, client Shangri-La Hotel in Singapore becomes the first licensed international restaurant.

to supply its many hotel and restaurant customers. This program could deliver the consistent quality Bruss wanted for its customers, so Pfaelzer requested an in-person meeting. Before he ever received a dime from the company, Berkowitz pulled interest from a high-quality distributor.

The Association Board permitted Colvin to hire Berkowitz for $2,000 a month for three months. "And by golly, he was worth it," Colvin says. "He had the keys to the doors of three top companies and their competitors." For those three months—February, March, and April 1982—the duo was on the road pitching Certified Angus Beef. During that time, Colvin also hired the brand's first official employee, Tim Hussman, who spent a lot of time with Berkowitz trying to convince distributors to join the program.

"My job was setting up the distribution network for foodservice and working with USDA regulatory. It was kind of a juggling act," Hussman says. "Leonard was very involved in helping me select the foodservice distributors. We would make the visit and he would show me what he was looking at. And I would show him what I was looking at and we'd compare notes. He was a great advisor."

Berkowitz's impact was seen almost instantly. He helped license Beef Specialists of Iowa by July 1982. The custom fabricator filled the smaller orders that large packers ignored, targeting high-end distributors with quality product. Slightly skeptical, the managers at the fabricators needed to see how they could make money. When Colvin promised the Iowa company ten cents a pound in premiums, "they were smiling all the way to the bank."

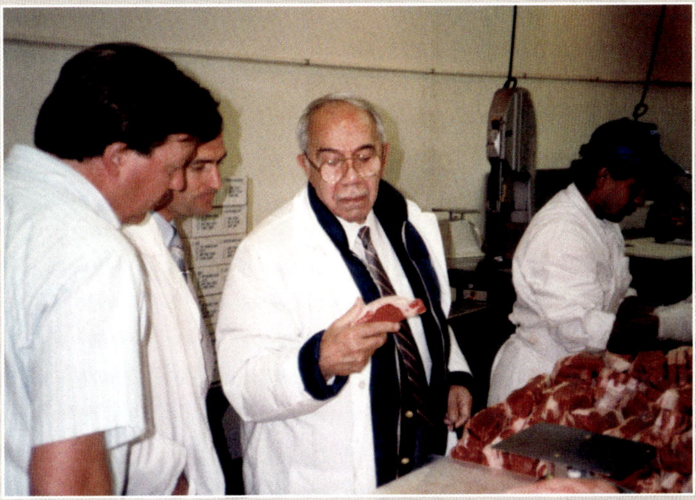

At a training session with Arizona-based City Meats & Provisions, Leonard Berkowitz (center) discusses a cut of beef. Berkowitz ultimately has keys to the doors Certified Angus Beef needs to enter. He proves to be an invaluable foodservice consultant for the brand.

The Consultant Who Almost Wasn't

When Leonard Berkowitz was asked to speak before the American Angus Association Board, he made a rather poor showing. He tried to tell the career Angus producers how to raise cattle, and tried to relate that to marbling. Needless to say, the message was not well received.

After the meeting, more than one member asked, "Why do we need this guy? We know more about beef than he'll ever dream about." But the Certified Angus Beef committee convinced the board that speech was just a mishap.

Dr. Bob VanStavern says, "Berkowitz came *that* close to not being involved in what we were doing."

Bud Matheson and Nancy Matheson-Burns license Dole & Bailey in 1983. The father-daughter duo introduces Certified Angus Beef to Massachusetts chefs and uses it as a way to differentiate the independent company from other distributors.

The business opened doors to several foodservice distributors who provided product to the world's best restaurants. While Beef Specialists was on board, it was still up to Colvin, Hussman, and the professor to sell the distributors. Berkowitz had a strategy: License Beef Specialists' distributors, which operated in the country's main markets. They flew to New York to meet with Marc Sarrazin Sr., owner of DeBragga and Spitler, in the city's famous Meat District. They chose New York City first because of its culinary significance: It's near to the world's premier Culinary Institute of America in Hyde Park and the finest restaurants. If Certified Angus Beef could succeed here, word-of-mouth advertising would spread like wildfire in the chef community.

According to Colvin, Berkowitz's opening line to Sarrazin was, "What if I can guarantee you the top levels of Choice, Yield Grade 3, and better?" That got the native Frenchman's attention. Then, Sarrazin, a gourmet butcher said, "Tell me about this program."

New York chefs, who were mostly European, respected Sarrazin because of his French background. If Certified Angus Beef representatives could convince Sarrazin of its virtues, the brand would soon be served at New York's finest clubs and restaurants.

"My father was very excited about the prospect because quality beef was hard to come by; it was very inconsistent," remembers Marc Sarrazin Jr., now company co-owner and then vice president of sales. "He was very anxious to jump on board."

When samples were shipped for Sarrazin to try, he was hoping to be impressed. "Hon-

CERTIFIED ANGUS BEEF®

QUALITY FIRST

For National Beef and CAB

by Jeri Lynn Gilleland

When National Beef Packing Co. joined the Certified Angus Beef (CAB) program last fall, it was almost as if CAB had found its Rosetta stone. Within 60 days of National Beef's induction to the program CAB total volume had doubled. That doubling tripled by the end of the first six months. Over one million pounds of CAB beef was sold during March, setting an all-time record since the program was introduced in 1978.

Mick Colvin, CAB director, explains, "National Beef gave us the volume and depth to service large distributors and retailers as well as the opportunity to make commitments to high volume distributors and retailers." Since CAB's inception, Colvin has been laying the foundation for a national beef retailing program. It's been a tough task but the program is starting to reap the benefits of his and the CAB committee's efforts.

For National Beef, the nation's sixth largest beef packer and, as they call themselves, "smallest of the big packers," the CAB program just coincided with their philosophy. John Jacobson Sr. heads the Liberal, Kan., firm with a philosophy rare to many in today's meat industry: Quality comes first. Quality minded customers have always been cultivated. Provisioners, restaurants, institutions and supermarkets looking for quality beef are National Beef's clientele.

Jacobson says, "We have never swerved from the top selection out of the feedlots. That has been our program since day 1. Whether the cattle are black or whether they are red or any kind of combination, we still have been concentrating on higher quality cattle. Our aim has always been and we continue to kill over 80 percent choice. That's our program."

Quality top concern

It was probably a mutual admiration that brought together CAB and National Beef. Colvin says, "We selected National Beef for their reputation in the food service industry. One of the things that impressed us about National Beef is that everyone there is quality conscious and quality oriented—both quality grade and quality control within the plant. Their interest in our breed of cattle was really encouraging."

CAB gave National Beef a chance to sell a product in which they believed. Jim Trail, vice president in charge of sales, says, "We were considering a selected meats program similar to CAB anyway. CAB came along and we thought the program would

Black hides have always been a favorite at National Beef Packing Co., Liberal, Kan. They are currently one of the five packing plants supplying CAB to distributors and retailers across the country. In addition to purchasing cattle from large feedlots, National Beef also slaughters from 125,000 to 135,000 animals a year from Supreme Feeders, Idle Wild's own feedlot near Liberal.

National Beef is "the smallest of the big packers." John Jacobson Sr. heads the packing plant as well as its parent company Idle Wild Foods Inc. Jacobson's philosophy, quality comes first, coincides with the CAB program.

ANGUS JOURNAL / June-July 1984

estly, the meat didn't look that much better. But when you cut it and ate it, the product was terrific. We signed on and started getting shipments from Beef Specialists of Iowa."

Specifically, the program's first foodservice distributor signed on July 12, 1982, and began selling its first shipments to well-known French restaurants like La Gaulyse, Le Cirque, and Le Reserve as well as the River Café, where celebrity chef Charlie Palmer was getting his start. DeBragga and Spitler also sold the branded product to the University Club, Rockefeller Center, and the New York Athletic Club, home of the NCAA College Football Heisman Trophy presentation.

"Certified Angus Beef helped our business enormously. It became a point of differentiation between us and everybody else," Sarrazin says. "We worked hard to market it and it became the single top-selling item. It still is today."

Shortly after the Sarrazins signed on, Colvin and Berkowitz visited Pfaelzer in Chicago. This time, the interested party stopped the presentation, shut his office door, and asked a question. "I was enthused by what they were saying, so I seriously asked: 'Can we have everything east of the Mississippi?' They both laughed," Pfaelzer says.

But he did land exclusivity in Illinois and started marketing to chefs in the Windy City, who were, of course, skeptical. "While the product gave us something our competitors didn't have, there was some confusion," Pfaelzer says. "We had a hard time convincing them it was upper Choice and Prime. But we sold them on the basis of consistency. It

National Beef
A U.S. Premium Beef Company

Licensing National Beef in fall 1983 allows the brand to make supply commitments to several high-volume distributors and retailers. Sales triple within six months, leading to the first million pounds sold. A summer 1994 article in the Angus Journal *details these milestones.*

1982

Leonard Berkowitz hired as foodservice consultant

Beef Specialists of Iowa inventories product

First foodservice distributor and first exports: DeBragga and Spitler

Association Board votes 8–7 to continue program

Ak-Sar-Ben Beef, Bruss Company, and Smith, Richardson & Conroy licensed as foodservice distributors; Bruss develops first box for cut steaks

Dugdale Packing (BeefAmerica) licensed

By 1984, supply channels reach from National Beef in Kansas to Caldwell Packing in Minnesota and Gartner-Harf Packing in Pennsylvania. Colvin, Leonard Berkowitz, and Tim Hussman work diligently to balance supply with growing demand.

was not a slam-dunk by any means. We worked very hard."

However, what seemed to be a homerun for both The Bruss Company and Certified Angus Beef was the international impact. Bruss had an existing client base in Singapore and sold this new product to the Shangri-La Hotel there. The Association saw that as a giant step for the brand.

"Singapore is a major crossroads of world commerce and the Shangri-La is rated by many as the best hotel in the world," wrote Richard "Dick" Spader, then executive vice president of the American Angus Association, in his "Lead In" for the August 1983 *Angus Journal*. "Certified Angus Beef is shipped regularly to hotels in Hong Kong, the Bahamas, and Puerto Rico."

DeBragga and Spitler also had great contacts overseas, specifically in Bermuda. After World War II, Sarrazin Sr. cut meat at the Castle Harbor Hotel there and met numerous up-and-comers in the island's foodservice business. When he took a job at DeBragga and Spitler and later purchased the company, his friends asked, "Why don't you sell us some meat?" As a result, Sarrazin Jr. says, "Bermuda became a very big piece of business for us. When Certified Angus Beef came along, it was just a matter of introducing the chefs to the product."

That made two foodservice distributors Berkowitz set up for the brand. By now, his month-to-month deal with the Association had expired, but the board permitted Colvin to hire him as a consultant for a year. In the later part of 1982, Berkowitz and Colvin signed two more foodservice distributors through Alan Kulakofsky at Ak-Sar-Ben Beef Co. in Omaha and Larry Dial at Smith Richardson & Conroy in Miami and Tampa. Things were definitely looking up. Certified Angus Beef operated its first month at a profit, and sold one million pounds in March 1983 while signing an influential foodservice distributor in the Boston area. Nancy Matheson-Burns, now owner of that

Tim Hussman (fourth from left) presents a plaque to the newly licensed Talk of the Town restaurant in Marietta, Ohio, in 1985. As Colvin's first employee, Hussman travels the country meeting with distributors and chefs. He helps put the brand on the map.

The First Employee Lost in the Big Apple

Tim Hussman, now the president of Newport Meat Company, a subsidiary of Sysco, was hired as the brand's assistant director in 1982. Growing up on a farm in Maryland, Hussman had never been to New York City. One day, he was sent to meet with DeBragga and Spitler.

After landing, he took a taxi to the Meat District and the cabbie said "I'm not going to pick you up. You have to find your own way back to the airport."

"I said, well, wait a minute, you can't leave me here," Hussman recalls.

Dealing with New York taxis was just one of the many learning experiences Hussman encountered during his thirty-month tenure with the brand. He traveled to forty states and was instrumental in signing the first distributors, including Newport, which eventually recruited and hired him. The experience, he says, "made my career."

Vito Aviolla, Gristede's in New York, shows off a tray of Certified Angus Beef steaks. The brochure is used to attract more retailers. Aviolla licenses forty-eight stores and advertises the brand in The New York Times.

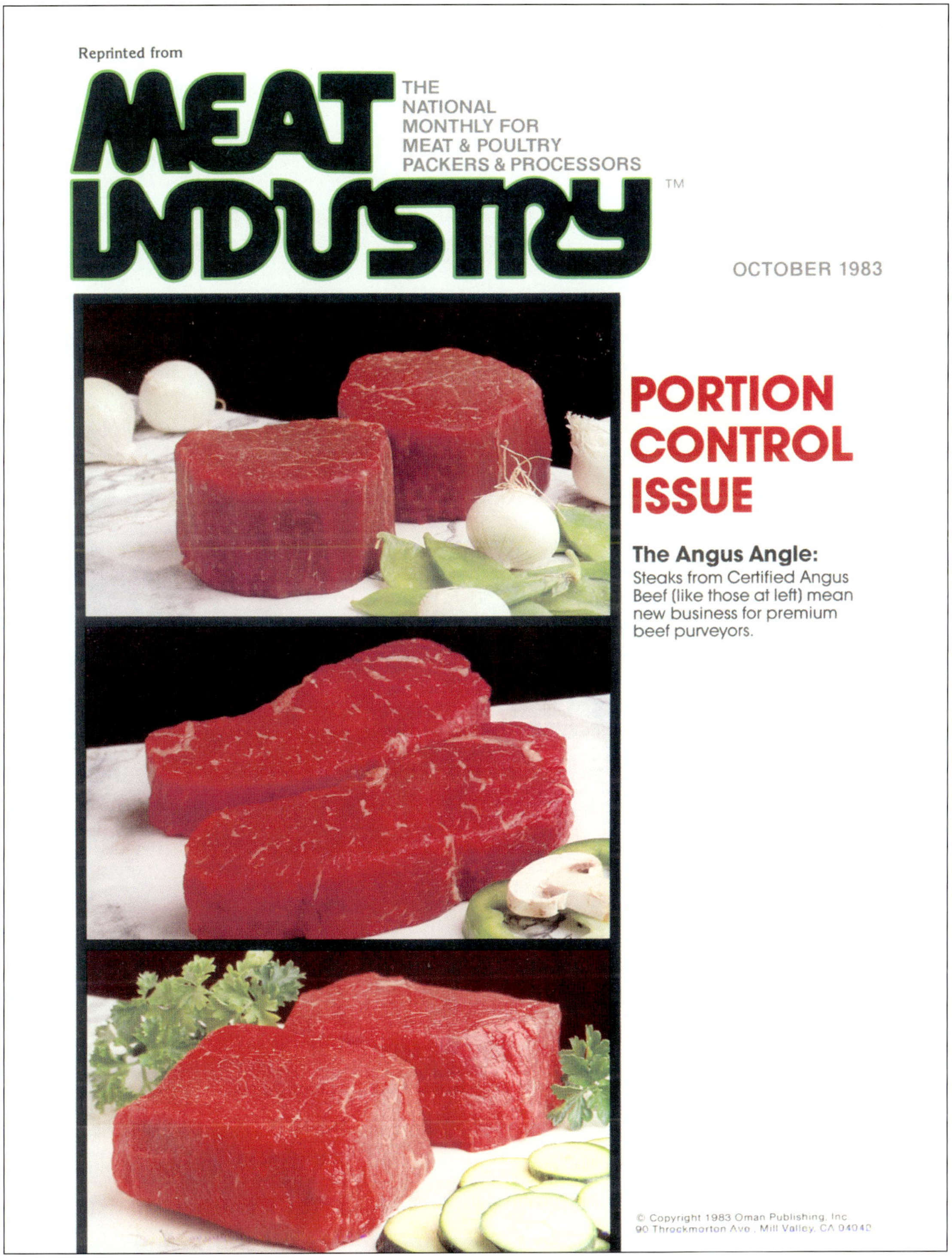

Reprinted from

MEAT INDUSTRY™

THE NATIONAL MONTHLY FOR MEAT & POULTRY PACKERS & PROCESSORS

OCTOBER 1983

PORTION CONTROL ISSUE

The Angus Angle:
Steaks from Certified Angus Beef (like those at left) mean new business for premium beef purveyors.

An article in the Meat Industry *magazine delivers one of those "we have arrived" moments. The "Angus Angle" features the brand with quotes from Colvin and Pfaelzer. After its release, calls pour into Colvin's office from interested distributors and chefs.*

In 1984, the Old San Francisco Steak House in Texas licenses four locations, making it the first chain account. Customers frequent the steakhouse, leading sales to 100,000 pounds of Certified Angus Beef a month.

company, Dole & Bailey, recalls listening to Colvin and Berkowitz.

"When they said cost was two dollars a pound more than conventional product, everybody glazed over except me," she says of their initial talks. "Some people thought nobody was going to be able to purchase this. But I saw an opportunity, an edge here for a chef to be committed; something different, something unique, something exclusive. While Mick was talking, I had a vision of a half dozen chefs using the product. Before the meeting was over, we made a verbal commitment."

In addition to licensing high-quality distributors, the program continued to garner great local, national, and trade press. But the business it was garnering was still relatively small when you looked at the beef industry as a whole. It made up a fraction of a fraction of total sales, Colvin says.

In order to grow, the staff had to license a large influential packer. But would a packer of surmountable size look at the company as competition or supplemental revenue?

The National Story

Berkowitz suggested they approach John Jacobson at National Beef Packing Company in Liberal, Kansas. The nation's sixth-largest packer was considering a selected-beef program. National bought mainly English-type cattle, most of which were black hided. Thousands of Oklahoma, Texas, and Kansas cattlemen—from three of the top five cattle-producing states—were within 160 miles of the plant on the Oklahoma-Kansas border. That meant access to supply, and the company had already been focusing on quality with more than 80 percent of its carcasses grading Choice or better. But Jacobson, then owner of National's parent, Idle Wild Foods Inc., recalls his retail and

Licensed partners meet in St. Joseph, Missouri, in 1984. They tour the American Angus Association and National Beef Packing in Liberal, Kansas. The meeting is a precursor to the "annual conference," which gathers customers from every segment—producers, distributors, chefs, and retailers—to share ideas and make their respective businesses better.

foodservice customers' craving good, consistent beef. Sure, most cattle were making the grade, but Jacobson's clients were not satisfied. "They were absolutely disgusted with the grading system," he says.

So, the solution was to provide his customers with upper Choice and Prime. In a 1984 *Angus Journal* story, National's vice president of sales Jim Trail said, "We were considering a selected-meats program similar to Certified Angus Beef anyway. This program came along and we thought it would have more credibility than our own."

In August 1983, National Beef Packing became the largest packer in the Certified Angus Beef system. It joined already-signed companies—Caldwell Packing (now PM Beef of Windom, Minnesota), Dugdale Packing in Iowa, and Union Packing Company of

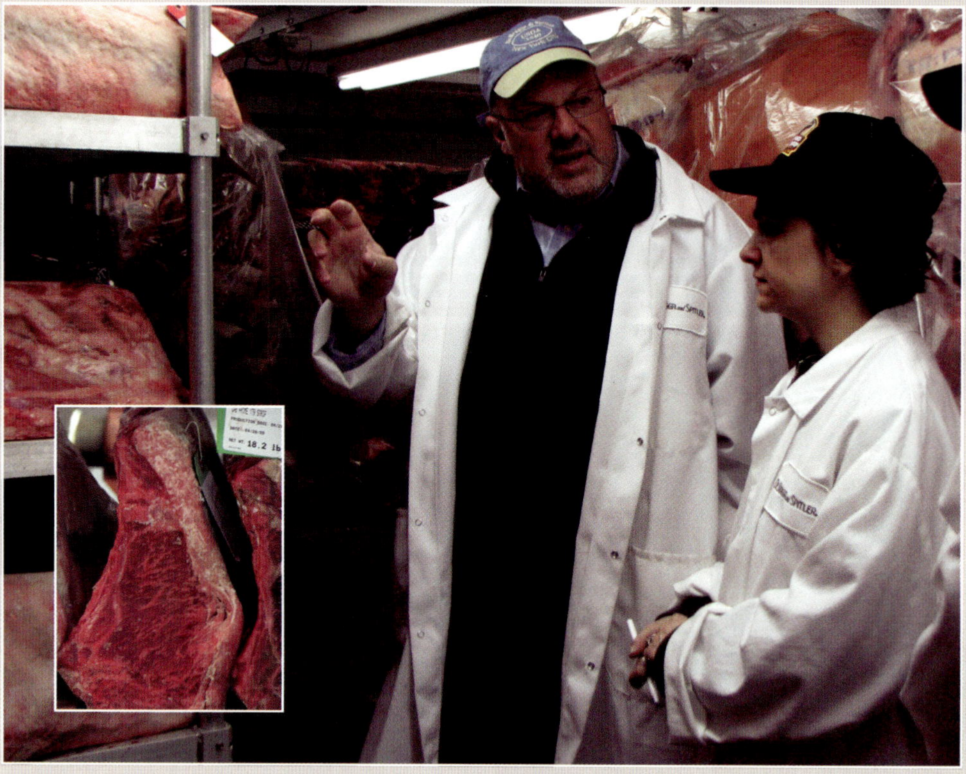

In 2009, Marc Sarrazin Jr. gives media a tour of DeBragga and Spitler's dry-aging room, where racks of Certified Angus Beef age under perfect conditions. Sarrazin meets with writers from Every Day with Rachael Ray, Food Arts, *and* Hampton Roads Magazine.

Dry Aging

In addition to keeping a deep Rolodex, Leonard Berkowitz knew how to make good beef great. A proponent of dry aging, Berkowitz insisted that carcass quality did not ensure appearance, tenderness, juiciness, and flavor on the plate. "The best of the best must be properly handled throughout all stages from kill to preparation and service," he wrote.

Aging made a recognizable contribution to final palatability. "Without aging, Certified Angus Beef product does not reflect its true worth."

Berkowitz contended that aging was the distributors' job and he worked with them to create the ideal dry-aging conditions: between thirty-four degrees and thirty-eight degrees Fahrenheit. "The higher the temperature [up to thirty-eight degrees], the more active the enzymes and the faster the aging. ... Aging stops when beef is frozen."

Just as aging affects fine cheeses and wine, dry aging enhances beef and gives it an intense, nutty flavor. Certified Angus Beef foodservice guidelines detail expectations that vary with the subprimal cut, of at least fourteen to twenty-eight days of dry or wet aging.

Omaha—which supplied numerous foodservice distributors. At the time, other retail-focused packers included Diggs Packing in Missouri, and Gartner and Harf in Pennsylvania. But these packers supplied a limited proximity. With National on board, the Certified Angus Beef program had greatly increased its potential for reaching consumers all over the country.

"We took a lot of pride in the program, because we took it from its inception all the way through today's success," Jacobson says. "It fit so nicely with us because we were always about quality."

Most of National's qualifying beef headed east. The largest single customer was Gristede's, a forty-eight-store chain of retail meat markets in New York City. Loins and ribs were in the most demand and for every primal cut, three cents per pound went back to the program. One cent per pound went back for chucks and rounds. Considering National harvested some 3,500 head of fattened cattle a day with 15 to 20 percent qualifying for Certified Angus Beef, it was making record sales every day. For Colvin, who had struggled for five years, there was now a reason for that smile on his face. Restaurateurs, packers, distributors, retailers, and ranchers were all finally seeing rewards. Could it get any better, he wondered?

Then, he received the October 1983 issue of *Meat Industry: The National Monthly for Meat & Poultry Packers & Processors*. On the front cover was, "The Angus Angle: Beef purveyors find a successful alternative for hard-to-get Prime." Unlike much of the trade press the program received, this story was neither a press release submitted by Evans nor written by an Association member. The *Meat Industry* editors saw a burgeoning program its readers could use to boost bottom lines, and its reporters objectively wrote about it. The story detailed distributors, National Beef Packing, Beef Specialists of Iowa, the certification process, and the demand for quality. Quoting Colvin, along with Pfaelzer and Charles Hunter of The Bruss Company, it was 140 column-inches of pure praise.

For anybody in the meat business trying to find quality beef, the story gave them a road map with no reason not to call Colvin, who was quoted in true salesman form. "A country club might want a little higher quality beef but doesn't see any advantage to signing an agreement or in promoting product," he said. "On the other hand, some other white-tablecloth restaurants do want to let their customers know. They sign an agreement, and we give them promotional materials, and they print that on their menus and describe why the beef is better, or promote [it] any way they want."

After industry leaders read the article, Colvin's phone rang. "And it never stopped."

The year 1983 seemed to begin a new era. The brand sold its first pound overseas and signed a large national packer that contributed to sales having reached one million pounds a month. It had its first profitable month and gathered great attention from trade magazines. But that was only the start.

As Ronald Reagan battled Walter Mondale for a second presidential term, the Certified Angus Beef brand was gracing the pages of *The New York Times* in the form of a full-page advertisement, 100-percent paid for by Gristede's. And farther south, the program signed its first licensed restaurant chain.

At four Old San Francisco Steak Houses in Texas, some hundred thousand pounds

1983

March sales exceed one million pounds, and first month sales exceed expenses

U.S. Patent & Trademark Office approves certification mark

First retailer beyond U.S. borders: Pueblo International, Puerto Rico

First major packer: National Beef

Colvin, Berkowitz, John Jacobson from National Beef, and four distributors meet in Kansas City

First international restaurant: Shangri-La, Singapore, licensed by Bruss

Meat Industry Magazine cover story leads to foodservice growth

Edd and Nina Hendee take Certified Angus Beef claims to another level. In 1984, their Taste of Texas begins offering "The $1,000 Steak Guarantee," which has yet to be cashed in.

The Taste of Texas Story

When Blue Ribbon Packing visited Taste of Texas, Edd Hendee was searching for an identity for his restaurant. Taste of Texas had evolved into a steakhouse, but Hendee questioned if it was hitting its potential. Then, Fred Pauley of Blue Ribbon told Hendee about Certified Angus Beef.

"I tasted it and after about the second bite knew that was exactly what I wanted," Hendee remembers. "It was exactly the flavor and the quality that we wanted."

Hendee scheduled a meeting with Pauley. He learned about the story of Certified Angus Beef and its scientific standards. Soon thereafter, Taste of Texas only sold Certified Angus Beef's cuts.

He also became loosely affiliated with some greater Houston-area Angus producers who heard about Taste of Texas' affiliation. Hendee says their enthusiasm and commitment to excellence was obvious.

"I thought this seems like an interesting synergy and partnership," he says. "Over the years, it has served us well, from the producer all the way to the customer. Certified Angus Beef has been the vehicle that we have used to not only provide the customer with a quality product, but also to differentiate ourselves in the marketplace."

As their restaurant becomes known for great-tasting steaks, the Hendees lay the chicken-fried steak to rest and vow it will never appear on their menu again.

Taste of Texas, which exceeded $12 million in sales in 2008, has helped Certified Angus Beef grow in Houston and around the country. Hendee, who is also a conservative talk show host on AM radio, convinced a Texas retailer to become a licensee by holding in-store taste tests at Randalls Food Markets and producing a video on backyard grilling. "We just had some fun with it," he says.

Hendee has fun with his $1,000 guarantee, too. In his restaurant, there's a poster that says: "$1,000 guarantee: If you can find a better steak with comparable price and size to our 16-oz Certified Angus Beef strip, this $1,000 CD can be yours."

He purchased the Certificate of Deposit in 1984 and it's still collecting interest.

"Our position has always been, if your steak isn't excellent, we're going to replace it immediately," Hendee says. But when somebody is really after that $1,000, Taste of Texas employees hand over a coupon for a steak dinner for two. If the steak is not perfect the next time? "We will line up three people of your choice and three people of my choice and we'll let them vote," Hendee relates. "Invariably, they think that the coupon was fair and they say, 'Oh, that was perfect.' "

After licensing his Taste of Texas restaurant in 1984, Edd Hendee becomes an outspoken supporter at conferences and distributor events. He frequently shares how the brand helps make the restaurant a destination synonymous with quality.

Opposite: In 1986, Litvak Meat Company hopes to increase supply by offering the first premiums for cattle qualifying for the brand. By December 1987, the USDA reports producers indeed receive more money for qualifying cattle.

of Certified Angus Beef products were served each month because of its consistency and quality. In operation since 1968, Old San Francisco always had quality product, but the low availability of Prime and inconsistency of Choice made it difficult to deliver its trademark superior dining experience. Because of the restaurant's great demand, Colvin gave Old San Francisco priority over other licensees.

"It was critical for us to establish this kind of program because quality meat was hard to get," said vice president Paul Sermas to the *Angus Journal* in 1984.

Like Gristedes, Old San Francisco paid for materials to promote the product, plastering the logo on its menu and taking out ads in papers. That was becoming a trend among licensees: They were marketing the brand themselves. Matheson-Burns painted the logo on Dole & Bailey's delivery trucks and made hats and aprons to give to chefs. At Oxford Trading, Robinson was knocking on retailer doors and switching coolers from regular Choice to the brand. But while distributors were doing the legwork, they were still amazed by Colvin's vision and follow-through.

"When we took a shipment, he would fly in every other time because he wanted to look at it and make sure you could identify it," Robinson says.

Matheson-Burns says that Colvin's greatest attribute was his ability to organize people and facilitate empowering meetings. These gatherings shaped the future. Although annual brand income exceeded expenses for the first time in 1984, many licensees say that was fueled by a meeting the year before in a rundown hotel room where seven people talked about their problems.

As Colvin saw the licensees marketing the brand like it was their own, he knew there was a great opportunity for retailers, restaurateurs, and distributors alike to talk about their issues and resolve them. That first meeting in 1983 included four distributors, Colvin, Jacobson, and Berkowitz. They talked about margins jeopardizing the future, the fundamentals of marketing the brand, and the importance of maintaining integrity. And it all happened at a wobbling table with two broken legs, Matheson-Burns remembers.

"That's when the fire caught," she recalls. "We were throwing any and all ideas, and answering: How can we make this work?"

As the program grew, the meetings increased to twenty-seven attendees in 1984, forty-three attendees in 1985, seventy-seven in 1986, and more than one hundred and fifty in 1987, according to a 1987 *Angus Journal* article. Those meetings were precursors to the annual conference, and became invaluable for the licensee as the only venue in which a packer could understand the producer's point of view, or a restaurateur appreciate a distributor's. At the time, there was no other across-the-board collaboration like it. Everybody from the farm to the plate was represented.

"We talked about how to solve problems," remembers VanStavern. "It wasn't what the packers' and distributors' roles were. We didn't have a distributor-versus-packer war but a moderated debate. They never became gripefests."

At many of the early meetings, Matheson-Burns was taken aback by the knowledge from the likes of Jacobson and Edd Hendee, who licensed his Taste of Texas steakhouse in 1984, the same year as Wormwood's Restaurant in Saco, Maine, licensed theirs. At the 1987 meeting, Hendee talked about the success of his Hous-

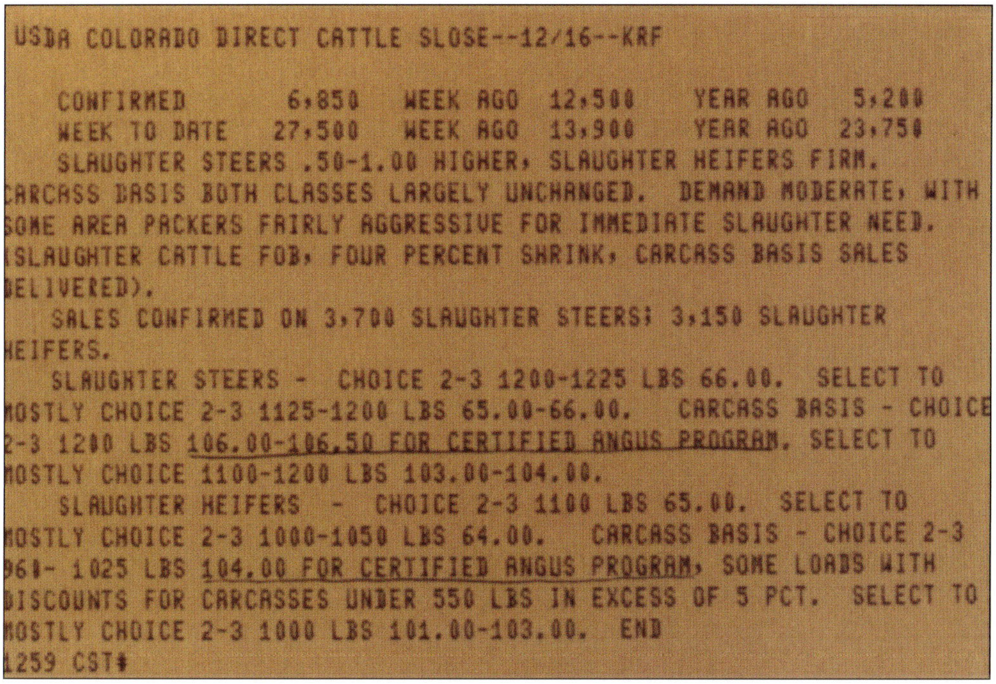
USDA COLORADO DIRECT CATTLE SLOSE--12/16--KRF

CONFIRMED 6,850 WEEK AGO 12,500 YEAR AGO 5,200
WEEK TO DATE 27,500 WEEK AGO 13,900 YEAR AGO 23,750
SLAUGHTER STEERS .50-1.00 HIGHER, SLAUGHTER HEIFERS FIRM.
CARCASS BASIS BOTH CLASSES LARGELY UNCHANGED. DEMAND MODERATE, WITH
SOME AREA PACKERS FAIRLY AGGRESSIVE FOR IMMEDIATE SLAUGHTER NEED.
(SLAUGHTER CATTLE FOB, FOUR PERCENT SHRINK, CARCASS BASIS SALES
DELIVERED).
SALES CONFIRMED ON 3,700 SLAUGHTER STEERS; 3,150 SLAUGHTER
HEIFERS.
SLAUGHTER STEERS - CHOICE 2-3 1200-1225 LBS 66.00. SELECT TO
MOSTLY CHOICE 2-3 1125-1200 LBS 65.00-66.00. CARCASS BASIS - CHOICE
2-3 1200 LBS 106.00-106.50 FOR CERTIFIED ANGUS PROGRAM. SELECT TO
MOSTLY CHOICE 1100-1200 LBS 103.00-104.00.
SLAUGHTER HEIFERS - CHOICE 2-3 1100 LBS 65.00. SELECT TO
MOSTLY CHOICE 2-3 1000-1050 LBS 64.00. CARCASS BASIS - CHOICE 2-3
960- 1025 LBS 104.00 FOR CERTIFIED ANGUS PROGRAM, SOME LOADS WITH
DISCOUNTS FOR CARCASSES UNDER 550 LBS IN EXCESS OF 5 PCT. SELECT TO
MOSTLY CHOICE 2-3 1000 LBS 101.00-103.00. END
1259 CST#

ton-based restaurant that earned some $1.6 million in sales and boasted a $100,000 advertising budget. He credited the Certified Angus Beef brand for helping his business become one of five hundred restaurants to receive the Exceptional Hospitality Award. Hendee criticized typical approaches to beef merchandizing and pointed out smokescreens like "USDA Good beef." He said, "I went into one place locally here and the waitress brought us each three menus—the light, the green, and the Cajun. Too many items. Too many dogs on the menu. When I worked for that restaurant, there were seven items on the menu and they were all good. Now there are sixty, and they're all mediocre. Mediocrity is available at the 7-Eleven for $1.99. Why go somewhere else for $14?"

A colorful speaker, Hendee made his fellow restaurateurs understand the need for consistency and quality while helping distributors see there's no substitute for the best. Hendee's speech was just one of many moments where licensees selflessly shared their stories. "The more stable Certified Angus Beef becomes, the better it is for us and for other restaurants that depend on a commodity-priced beef brand," Hendee says.

Matheson-Burns says everybody learned from each other. "That's how this brand really took off."

As the brand grew, the demand for Angus bulls increased. By 1986, one million Angus-influenced cattle were evaluated for the brand per year and Litvak Meat Company started offering incentives for cattle qualifying for the program. In a letter to Colorado cattlemen, Litvak officials wrote: "Since this is the time of year to start replacing cattle for winter and spring markets, we want to again make you aware of our earnest effort to slaughter as many Black Angus heifers and light-weight steers as possible. Please consider buying the Black Angus breed as you purchase your fall feeder needs."

There were developments happening on the processing side, too. In October 1986,

1984

First restaurant chain: Old San Francisco Steak House

Gristede's licenses 48 stores and places ad in *The New York Times*

Taste of Texas and Wormwood's Restaurant licensed; later longest continuous restaurants

First annual conference: 27 attend and tour National Beef

Program earns $281,276.44, the total invested by Association to date

Program office moves to new West Salem, Ohio, location

Carcass data services become another avenue for channeling supply. Certified Angus Beef board member Dick Janssen works with Colvin to analyze data. He is pictured with board members (from left) Bob Schultz, Fred Smalstig, T.D. Steele, Dr. Bob Hillier, (Janssen), and Dick Spader, Association executive vice president.

National became the first packer to imprint the logo on a Cryovac vacuum-packaged bag. Festival Sausage pioneered value-added products, using Certified Angus Beef chuck meat to formulate summer sausage, salami, and beef sticks. Perhaps most amazing was that a little not-for-profit company was leading the packing and cattle industry in so many respects.

A 1987 *Crittenden Cattle Buyers* newsletter published in Novato, California, explained: "Years of mistrust between packers and producers foul attempts to control consistency and quality in beef production. Yet the success of the Certified Angus Beef program proves that trust can provide rewards for all. The program brings together packers, feeders, and producers as never before. Outsiders criticize the program for not offering premiums. ... Yet producers reap rewards because the breed is being promoted and demand grows. That pushes up prices for Angus cattle."

The evidence for that push in prices was published on December 16, 1987, when USDA detailed the Certified Angus Beef premium in its *Colorado Direct Cattle Close*. "That was the first time a premium had ever been paid. Until that point, packers were sorting off carcasses and getting a premium but not offering anything back to producers. Of course, the brand was created to send money back to producers. So, that was a very significant step," says John Stowell, meat director for Dole & Bailey and former Certified Angus Beef employee.

Stowell was the program's first Supply Development director. In 1988, he embarked

Nancy Matheson-Burns is an innovative marketer. When she learns the 1988 U.S. Culinary Olympic team is training in her backyard, the young Dole & Bailey executive supplies Certified Angus Beef and makes "Len," a two-year-old Angus bull, the team mascot.

The Bull Named Len

While a group of American chefs was training for the 1988 Culinary Olympics, they had a team mascot—a two-year-old Angus bull named Len, after Leonard Berkowitz. Len was owned by Nancy Matheson-Burns, whose Dole & Bailey sponsored the team by supplying Certified Angus Beef cuts for the cooking.

"I donated a lot of product to get it in their hands, and we spent a lot of time with those guys when they were cooking during team practices," she says.

At the practices, Len would hang out in the pen and people would take pictures next to him. Matheson-Burns says the people just loved him.

"Certified Angus Beef product was getting a lot of attention in the community because of the Culinary Olympics," she says.

Recognized as the first Certified Angus Beef "Pioneer," Oxford Trading Company's Charlie Robinson remains as supportive today. Robinson proudly displays an Angus bull statue on the company's roof.
Photo credit: Fred Minnick

Leading the Herd

Taking the first step to partner with the brand seems logical today, but back in the late 1970s and early '80s it wasn't quite proven. Companies were taking a leap of faith in Colvin and the value the brand would deliver to their businesses. Because of several early pioneers who truly embraced the brand and made it their own, it flourishes for the benefit of more than fifteen thousand licensed businesses worldwide and even more Angus cattle producers.

Throughout the years, a number of people have been recognized for initiating the brand's early success. The first Angus bull statuette was presented to Charlie Robinson, Oxford Trading Company, in 1986. Equal recognition honored Dr. Bob VanStavern in 1987, Fred Johnson in 1988, Leonard Berkowitz in 1989, and Mick Colvin in 1990.

Additional honorees and their point of influence included:

1987: Steve Pearson, Beef Specialists of Iowa

1988: Vito Aviolla, Gristede's

1989: Charlie Vignieri, Kenosha Beef International

1990: Jim Addison, National Beef and Nebraska Beef

1990: Bob Facciola, Facciola Meat Company

2002: C.K. Allen, Executive Vice President, American Angus Association

2009: Edd Hendee, Taste of Texas Restaurant

on an invaluable mission—to improve the supply of quality cattle that would meet the brand's specifications. Stowell, a former Litvak employee, was selected to head the program because he was handpicking cattle for Litvak that he knew would meet the specs. Stowell had a knack for selecting live animals and carcasses.

"The program was growing so well, but they were not getting any more cattle meeting the specifications," he says. "There was a lot of genetic ability for the right type of cattle, but getting those cattle identified and in the right feedlot was a challenge. The idea was to channel the proper genetics and feeding programs into the Certified Angus Beef program."

Richard "Dick" Janssen, a Certified Angus Beef Board member in 1987, was influential in setting up the supply development division and starting the Carcass Data Collection Service. A long-time Association member, Janssen had watched the program grow from its infancy through to revenue generation and was amazed at how it was appreciated by chefs and distributors. "I thought, what we needed was to try to produce this on a more regular basis," he says. "Normally, when you decide to increase a product, you go build it. But what we were doing was sorting what's on the rail."

During a trip with Colvin to the Liberal, Kansas, National packing plant, Janssen saw the carcasses move via rail and then analyzed data between the poorest and best. Janssen saw the data as similar to that used in the Association's expected progeny differences (EPD) genetic selection tools. "We thought, since we know how EPDs work, we should be able to take this carcass data and give it back to the producer, so we could produce more Certified Angus Beef cattle," he says. "That really made a difference in the breed."

For Janssen, the success of the brand was certainly welcomed. The owner of Green Garden Angus, Lorraine, Kansas, had seen so much decline in his cattle breed's popularity over the years, it was nice to see a positive. In fact, things were so dismal that, Janssen says, the Association told members that the breed had reached its maturity and registration numbers were expected to drop to under a hundred thousand. "We were cutting costs like there was no tomorrow. But there were some of us thinking if we do this thing right, we will be the Holstein of the beef industry."

And Janssen was right. As the 1980s rounded out, the brand had signed one of the "Big Three" packers. It had international accounts, and *The Art of Cooking with Certified Angus Beef* cookbook could be found on the bookshelves of chefs across the country. After more than ten years of operation, the revenue was just flowing in, and then back out to promote the growing brand.

"We had to create new and bigger budgets every quarter," Janssen remembers, "because we just kept growing." Along with that growth came the need to place more emphasis on product tracking.

1985

First retail chain on West Coast: Olson's, later purchased and uses Quality Food Centers name

First computer purchased to enhance product monitoring

First processor: David Berg & Company, corned beef brisket

First export-focused company: Midamar Corporation

Three more packing plants licensed, including Litvak Meat Company

The brand launches in retail stores initially, and those sales grow along with restaurant business for several years. Retail sales really bloom in the late 1980s as major packing companies are licensed.

4 Built on Solid Structure

When Mark McCully, Certified Angus Beef assistant vice president for supply development, travels for work, he almost always strikes up a conversation with his fellow airline passengers. A student of business philosophies and strategies, McCully listens to people talk enthusiastically about their professions. He learns about new products and strategies, improving his understanding of the overall business world. But when it's his turn to talk, it's almost guaranteed that the other person will not comprehend the structure of Certified Angus Beef LLC.

Although other businessmen recognize the company logo and most likely have enjoyed a tender, flavorful cut of the product, it's hard for them to fathom that the organization doesn't own cattle, packinghouses, packaging facilities, restaurants, or retail outlets. And when McCully tells them that it's a not-for-profit company and the only thing it owns is the logo? "They get even more confused but really fascinated with our business model. I have never run across somebody that says, 'Your business model is just like Company X.' Nothing else like our company exists and it's hard for people to understand."

The concept of relying so heavily on packers and other licensees is so alien that some consultants have advised shutting down Certified Angus Beef before it falls apart, because "there is no way it could work," McCully says.

The structure of the company, while not simple, is not precarious, as some outsiders might believe. It begins in the feedlot or ranch where Angus-influenced cattle are purchased by licensed Certified Angus Beef packers. After the producer delivers the cattle,

Certified Angus Beef operates under a business model that cannot be found among other companies, and can be hard to explain to those unfamiliar with the meat industry. This flow chart illustrates the company's unique structure from Angus ranchers to consumers.

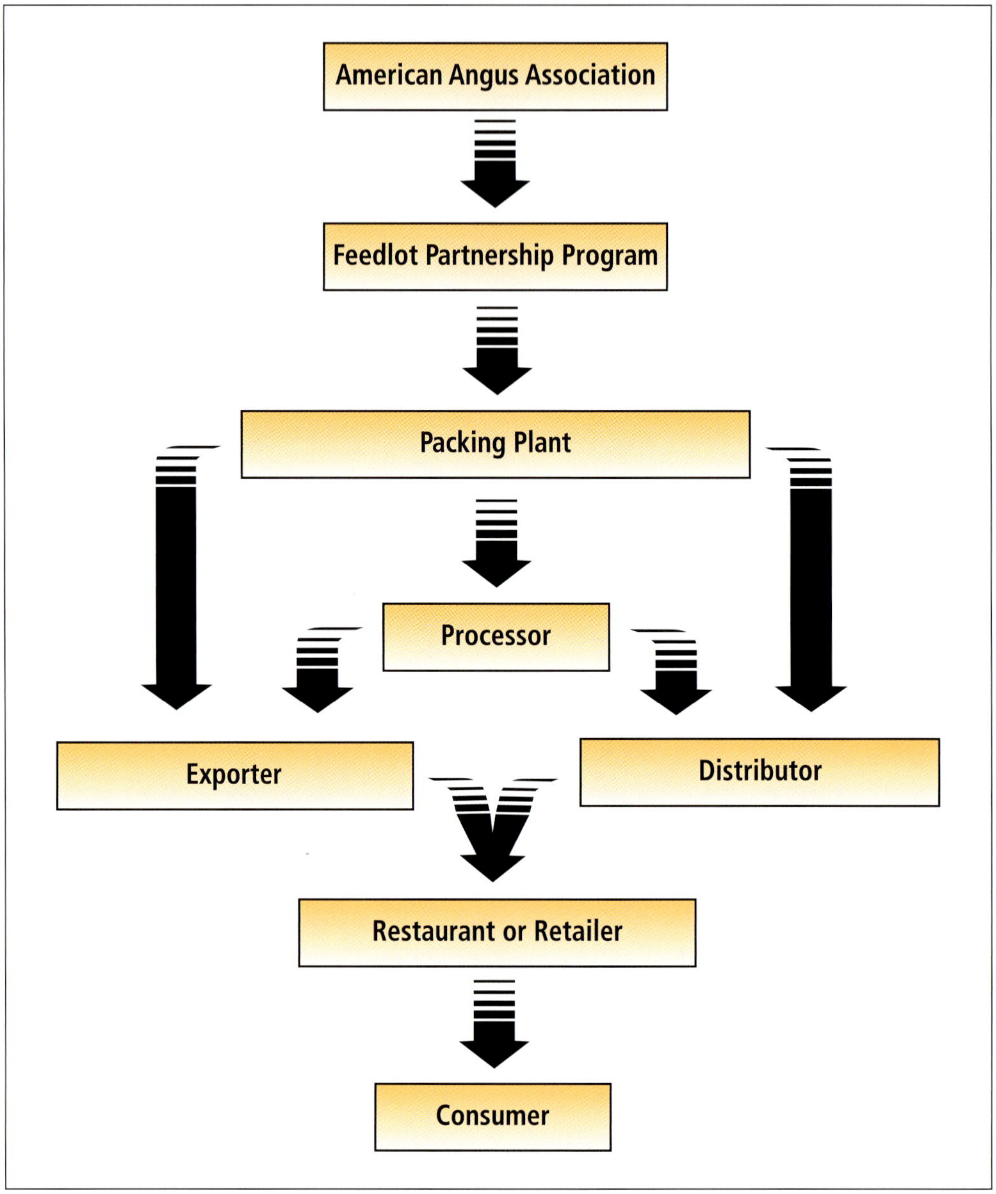

a trained plant worker identifies the black or AngusSource® cattle as they are harvested and their carcasses hung on a rail; and those that have a black hide or AngusSource tag are marked. A day later in the cooler, the government grader judges the ribeye area and other attributes. If they meet the maturity, marbling, consistent sizing, and other brand specifications, he will give it the Certified Angus Beef mark. The packer, who pays a premium when he purchases the cattle, then cuts up the carcass to sell as boxed beef to licensed Certified Angus Beef distributors, and pays a commission to the brand of two cents per pound on average.

When the distributors receive the products, they put their company's finishing touches on them. For example, all foodservice distributors further age the meat fourteen to twenty-eight days; some feature dry aging and cut to customer specifications. Meanwhile, retail distributors send boxed beef to the retailer, who cuts the items into

The First Office

While the structure of today's Certified Angus Beef might be difficult for outsiders to comprehend, its humble beginnings are mind-boggling. Up until 1984, when the company opened its first office in West Salem, Ohio, employees operated out of Colvin's home.

When Tim Hussman interviewed for a job in 1983, he remembers pulling into Colvin's driveway. "I said this is nice. Are we going to go to your office?" Colvin gave one of his famous grins and said, "This is my office."

In 1994, the company moved to a new facility in Wooster, Ohio, where three years later it doubled in size. It now has room for one hundred employees and meetings with Angus producers from across the country and customers from around the world.

For the first six years, Colvin and three staff operate the brand from a one-room office in his home.

The Association Relationship

Deeply rooted in its ties to the Association, Certified Angus Beef was structured exactly like its parent company.

"We were able to ride on the Association's coattails with our mission being so closely aligned to their not-for-profit mission of growing the demand for Angus cattle through the membership organization," says Brent Eichar, the brand's senior vice president.

Up until 1989, Certified Angus Beef sent all its revenue and accounting records to the Association. In a sense, the Association was overseeing every financial aspect of its quickly growing subsidiary and redistributing the funds back to Certified Angus Beef.

But at the September 13, 1984, Association Board meeting, Johnson took the first step toward autonomy. He presented a recommendation for the restructure of the brand as its own entity. Harvey Lemmon, from Georgia, led the vote in November during the Association's annual meeting and became chairman of the Certified Angus Beef Board starting in January 1985. Fellow board member T.D. Steele, from Virginia, moved to the Association Board in 1988 and helped complete the transition to autonomy. By 1989, the Association only audited its subsidiary's books once a year.

For Colvin, that was an important step. "We needed to be investing everything we could in Certified Angus Beef to make it grow and profit our licensees, rather than depending on the Association to set those numbers," Colvin says. "We had a separate accounting structure then and could focus on building the brand."

At Colvin's retirement in 1999, the staff presents him with this Conestoga wagon wheel signed by them. It represents one of their most cherished "Mickisms."

Mick's Conestoga Wagon Wheel

(from a 1999 article in *Angus Journal*)

Staff around the Certified Angus Beef headquarters call them "Mickisms," favorite sayings and stories Mick Colvin uses:

- "You can't un-ring a bell."
- "Walk a mile in someone else's moccasins."
- "We didn't know it couldn't be done."
- "Stick with the girl who brought you to the dance."
- "I've never seen our staff stronger than it is right now."
- "We've never been in a better position than we are now."

One Mickism, though—the Conestoga wagon wheel—isn't a simple sentence, but rather an analogy of how he believes the program works and what has enabled it to get where it is today.

"Everyone wanted to be a pioneer until they were asked to take that first step into the Conestoga wagon as part of a wagon train headed west to unknown territory," Colvin said in a 1995 Christmas-party talk to his staff. "No one wanted to do that." Much like the canvas-covered wagons that forged ahead to new frontiers, this branded beef program carried the Angus breed into unknown and unexplored market areas.

The Certified Angus Beef program is, he explained, "symbolic of a wheel on the Conestoga wagon. The axle is our consumers because everything revolves around them; and the hub is our licensees. Each spoke is a division of the program. Employees support the licensees. The outside steel rim, which holds everything together, is the program's mission statement and the essence of the program itself."

After his presentation, Colvin, feeling a bit proud of what he thought had been a perfect comparison, was taken aback by an innocent question from a younger audience member. "Someone asked, 'What's a Conestoga wagon?'" Colvin recalls with a chuckle.

typical steaks, roasts, and grinds. Some chucks and rounds are sent to processors that create specialty items such as meatloaf, frankfurters, hamburger patties, and deli meats. Although licensed packers and processors pay a few pennies per pound, the licensed distributors do not pay the company one cent. For that matter, neither do the licensed restaurants and retailers. Nonetheless, Certified Angus Beef provides these companies a plethora of point-of-sale marketing material, advertising, literature, use of the logo, and most importantly, the assured value of taste.

The company executes many strategies to improve the supply of Angus-influenced cattle and increase the premiums producers receive. It also initiates tactics to encourage every licensee's potential for success. But the only way Certified Angus Beef makes money is through the cents-per-pound premiums paid by packers and processors.

To put that structure into perspective, let's look at another food company that has had a great impact upon another commodity. If Starbucks had the same business model as Certified Angus Beef, it would not own coffeehouses but instead license bean processors who would then sell to distributors who would sell to retailers and foodservice outlets who would sell to consumers. Starbucks would work with coffee growers, helping them with genetics and other management practices in an effort to improve the farmers' margins. Starbucks would receive a modest premium from the processors with the goal of improving the livelihood of coffee growers and the demand for high-quality java.

Could anybody really see Starbucks working under such a model? While it has done more for international coffee growers than any other organization, Starbucks is a publicly traded company that generated $9.8 billion in revenue in 2009. Some of that money was used to pay top executives multimillion-dollar salaries, like the board members from other Fortune 500 companies.

To the contrary, Certified Angus Beef's nine-member board is comprised of eight unpaid Angus producers and one unpaid meat industry professional, and its staff receives modest pay. Originally, the board was comprised of five members selected by a vote of the American Angus Association. In 1996, the Association Board voted to increase the size of the Certified Angus Beef Board to nine, including one or two from the meat industry. Their responsibilities are to vote on "big picture" items such as new presidents or the yield-grade specification change of 2006. They also oversee the company's financial status and long-range planning. However, they do not intervene in day-to-day operations, such as marketing, supply development, or account management.

Over the years, there have been a few meetings with misunderstandings and heated debates, inevitable for a board of this stature, yet some past members say it was the highlight of their career.

"I've sold a lot of cattle on the grid because of Certified Angus Beef," says Henry Bergfeld, chairman of the branded beef company's board in 1997–98 and Fred Johnson's former farm manager. "Certified Angus Beef enhanced some of what I was doing and added big value to my life."

The company's employees like to hear such statements because they reflect the mission statement that was formalized in 1997 and is posted prominently on the wall just

CERTIFIED "The ONLY way to be Sure" ANGUS BEEF

Mission Statement:
To increase demand for registered Angus cattle through a specification-based, branded beef program to identify consistent, high-quality beef with superior taste.

1986

Festival Sausage licensed for summer sausage and salami

Contest encourages producers to contact local restaurateurs; Angus breeder Joel Stivers, New Jersey, wins

First restaurant in Bermuda: Hog Penny Pub; grocery stores include White & Sons and Supermart Fine Foods

Certified Angus Feeder Program begins identifying high-quality Angus cattle for sale

Foodservice distributors number 35, including Pacific World Enterprises (later Triple J Five-Star Wholesale Foods) and Provimi de Puerto Rico

Litvak Meat Company offers first packer premiums for qualifying cattle

One million Angus-influenced cattle identified during year; Dubuque Packing (later BeefAmerica) licensed

inside the front door of the office: *To increase demand for registered Angus cattle through a specification-based, branded beef program to identify consistent, high quality beef with superior taste.* In fact, that mission is stressed during job interviews.

"We have always told people we interview that if you're here for the money, this is not for you," says Brent Eichar, senior vice president for Certified Angus Beef, whom Colvin hired in 1987 before Eichar graduated from The Ohio State University. "It has always been about our mission statement."

In light of the enduring belief in that mission, it's no wonder Angus producers are enthusiastic supporters of the brand. From board to staff, follow-through on the mission has kept the company's integrity intact; it's always shared with new licensees as well.

The process for selecting a new licensee is a simple one. In foodservice, licensed distributors often approach restaurants they think would benefit from becoming licensed. If the restaurateur agrees, they fill out the short and sweet licensee agreement and adhere to a set of simple rules. For example, if the restaurant signs up to carry Certified Angus Beef strip steaks, it acknowledges it will not carry a Choice or lower quality strip steak. As for licensing a retailer, Certified Angus Beef account managers meet with interested parties and thoroughly go over the case display and labeling policies.

Andy Malcolm's foodservice distribution company, Malcolm Meats near Toledo, Ohio, was licensed in 1988; from the beginning, Malcolm believed in the program's potential and opportunity. But the way the organization treated all licensees as equals especially struck a chord with him.

"They expect you to represent that brand in the most professional way, and anyone who would ever do anything that wasn't at the highest level of integrity would be out of the program," says Malcolm, who is now a vice president for Sysco Corporation and chairman of Sysco Specialty Meat Companies.

Indeed, no licensee receives special treatment, even if they're the world's largest meat packing company.

"From the very early days, working with Mick Colvin, all the way through to today, the Certified Angus Beef team strongly defends the brand—with an energy and enthusiasm that's outstanding—to maintain the reputation that the brand enjoys," says Charlie Mostek, executive vice president of sales and marketing for Tyson Fresh Meats. "They take a great deal of pride in the production of the product and understand the value of it, both to the cattlemen and to the packing plant."

Licensing Packers

At the core of the company's foundation are its relationships with the packing plants. These were forged early on, when Colvin was pitching the program to packers in the 1980s. He faced many challenges as the program was still relatively new. Some packers didn't necessarily like the idea of separating the good meat from lesser performing

Outsiders and Their Ideas

Over the years, consultants and industry professionals have had advice for Certified Angus Beef about how to "improve" its business model, going for-profit or even separating from the American Angus Association.

Many retailers and restaurateurs, not willing to pay the premium, have tried to get Certified Angus Beef to lower its marbling standards to get a less expensive—and lower quality—product.

Large marketing firms have encouraged the company to focus more on national campaigns and less on licensee-oriented strategies in their local markets.

These unique suggestions are not isolated to consultants. One former employee argued that the black steer should be removed from the logo because people don't want a connection to the farm when shopping for food. But the company stood by Colvin's observation: "We always want to be connected to the farm."

The Certified Angus Beef staff often weighs ideas from within and without, promoting open debate, tempered with the collective wisdom born of experience with its unique business model. Typically, outside advice has not struck a chord because it lacks that insider insight.

People—both licensees and staff—have always been key to the brand's success. With Mick Colvin at the helm in 1989, his crew is (left to right, first row): Teara Blessing, Norma Noletti, Sharon Swanson, Vickie Catteau, Virginia Colvin, Linda Dickens, and John Stowell; (second row) Glen Ledall, Alan Waggoner, Mary Ferguson, Colvin, and Roger Clark; (third row) Brent Eichar, Kirke McKean, and Bill Wylie.

carcasses, while others didn't want to pay Certified Angus Beef a premium. But Colvin and his few employees never let up, preaching the scientific standards.

Excel, licensed in 1988—now Cargill Meat Solutions (CMS)—was the first of the "Big Three" packers to take the chance.

"Bringing on a USDA-certified program was a new undertaking," says Bill Buckner, Cargill CEO. "At the time, CMS was solely USDA quality grades: USDA Prime, USDA Choice, USDA Select, and No-Roll. Due to the nature of Certified Angus Beef being the first third-party commercial boxed-beef branding effort, it took quite a bit of discussion and work with procurement-operations-sales to coordinate and align the supply chain to successfully deliver on the promises made."

After signing Excel, the packing world took notice and Colvin went after IBP (formerly known as Iowa Beef Processors and now known as Tyson Fresh Meats).

When Colvin met with Mostek and Dick Bond, who until January 2009 was the CEO at Tyson, he met a brick wall. Mostek remembers Colvin talking about those scientific specifications "for the program that nobody had ever heard of, and we kind of looked at him with a jaundiced eye and said, "You know, we're really about our brand, not somebody else's. See ya!' "

As Certified Angus Beef was growing, IBP began to appreciate the brand's specifications, while Excel and National Beef had shown the packing industry that the burgeoning Certified Angus Beef program was worth the risk.

"IBP recognized the value of the brand. Bond and Colvin laid out the plan for IBP entering the Certified Angus Beef production arena, but also a plan for growth. Because at that time, I would venture to say that there weren't as many plants certifying Certified Angus Beef as IBP had in production capacity," Mostek says of IBP, which was acquired by Tyson in 2001 for $3.2 billion. "Mick was looking at it as an opportunity to almost double the business. It took a few years to get it 100-percent developed, but Colvin's persistence paid off."

IBP signed on in 1992, and by 1995 the company had eleven plants producing Certified Angus Beef, accounting for one-third of the licensed processors. To many, the IBP partnership was the "we have arrived" moment because this was the world's largest beef packer and the most influential. Among other innovations, IBP introduced boxed beef in 1967 and experimented with single-story buildings to enhance automated operations. The company used the same tenacious strategies with the Certified Angus Beef program.

"IBP was huge because it added so much supply and opportunity to move overall market share up," says Alan Waggoner, former packing director and assistant vice president for Certified Angus Beef, who helped Colvin bring IBP on board.

In helping get the IBP program started, Waggoner worked closely with Jerry Holbrook, who was pulled off of central sales to concentrate solely on Certified Angus Beef. Holbrook says the two companies worked closely together and the partnership helped IBP establish this value-added category.

Although some people in the beginning of its development said Certified Angus Beef would never produce hamburger meat, those opinions softened when IBP offered ground chuck, sirloin, and round. Within two years of licensing, IBP was also providing customers a full line of quarter-inch trimmed product, trying to maximize the use of the whole carcass. In various articles during that time, IBP officials said the value-added programs were increasing demand for Certified Angus Beef and Angus cattle. "In those days, we had a lot of trouble getting enough pounds per head to justify the expenses associated with it," Mostek says.

While IBP was gearing up its ground-beef program, Colvin was working on another packer, Monfort Inc. Like IBP, Monfort staff had concerns about the program and didn't know how its customers would receive the brand. But Mike Croot says Certified Angus Beef executives were really convincing.

"They laid out a well-drafted plan and had their ducks in a row. That was the one thing we talked about; we knew they were going to stand for quality," says Croot, vice president of sales for JBS Swift & Company, a successor to Monfort. "They weren't just taking any black animal. The animals had to meet certain criteria and everybody felt good that it was going to be a great eating experience."

Once convinced on the product, Croot, Dick Monfort, and others debated on whether they wanted to pay commissions to Certified Angus Beef. They asked: Is it going to be too much? "Everybody kicked it around and just said, 'Hey, all we can do is just give it a run and see how things turn out,'" Croot says.

JBS Swift, then as Monfort, became a Certified Angus Beef licensee in July 1995 and has enjoyed a long-standing relationship.

EXCEL® ibp® Monfort

Certified Angus Beef signs the three largest packers from 1988–95, growing access to 70 percent of the nation's fed cattle. Over the years, great relationships have developed between major packinghouses and the brand.

Today's Ten Science-Based Specifications

The brand's specifications are essentially the same as those VanStavern identified in 1978 to align with scientific research and ultimately ensure acceptable eating quality. In the early 1990s, USDA evaluation procedures required clarification in how to limit the influence of dairy and Brahman-type cattle. In 2006, the Certified Angus Beef Board led changes to accept genetically identified AngusSource cattle, and set three specific parameters for yield to maximize consistency.

After meeting the live-animal prerequisite of at least 51-percent black hide or AngusSource enrollment, cattle are evaluated on these specifications:

Marbling and maturity

1. Modest or higher marbling—for the taste that ensures customer satisfaction
2. Medium or fine marbling texture—the white "flecks of flavor" in the beef that ensure consistent flavor and juiciness in every bite
3. Only "A" maturity—the youngest classification for superior beef, color, texture, and tenderness

The next three specifications ensure uniform, consistent steak size:

4. 10- to 16-square-inch ribeye area
5. Less than 1,000-pound hot carcass weight
6. Less than 1-inch fat thickness (typically processed to ¼-inch trim or less)

Photos courtesy of the National Cattlemen's Beef Association, and the U.S. Department of Agriculture, © NCBA

And finally, four specifications further ensure the quality appearance and tenderness of the branded product:

7. Superior muscling—restricts influence of dairy cattle
8. Practically free of capillary ruptures—ensures the most visually appealing steak
9. No dark cutters—ensures the most visually appealing steak
10. No neck hump exceeding 2 inches—safeguards against cattle with more variability in tenderness

In comparison, USDA quality grades for beef are primarily based on just two factors: marbling amount and maturity.

Retail growth continues into the 1990s, proving that quality is important for at-home meals. Major grocery chains from Pennsylvania to Wisconsin to Texas license, leading the number of stores beyond 1,000 nationwide in 1993. Retailers advertise the brand's superior quality.

"People trust the brand," Croot says.

Over the years, Certified Angus Beef's licensed packers have found that paying the small premium to the company to fund promotion and program development has paid dividends.

"We're utilizing over three hundred pounds of beef per head as Certified Angus Beef product," Mostek says. "It offered us a marketing program with real results to help us move more beef further up the value chain. When Certified Angus Beef came by with a fairly simple, but very strict set of parameters for certification, it became very quickly apparent that we had a really high-quality product. And I think for the business in general, what we were able to do is bring beef that aged almost as well as Prime at a far less costly process, and in more adequate supplies. With Prime running generally in that 1- to 3-percent range of the total mix, it would just be cost prohibitive to try to develop a really large amount of Prime. In the case of Certified Angus Beef, we were able to develop a ready availability, and it really helped steak chains and retailers that needed a premium beef program."

1987

Andy Angus mascot begins appearing at food shows and store openings

Art Culinaire features brand with Chef Jean Banchet

Record 24.2 percent of identified Angus-influenced cattle in licensed packing plants qualify for brand

Testing Integrity

Very early on, Colvin and his cohorts realized that a man's word wasn't enough to ensure companies held up their end of the license agreement. There was no employee physically watching the packinghouses or restaurants to make sure they were actually selling Certified Angus Beef product with the proper logo. In fact, there was great potential for fraud here: A licensee could easily take Select or lower Choice product and place a Certified Angus Beef steak-pick on the meat, and the consumer might not know the difference. That's why Colvin and assistant executive director Mary Ferguson sorted through packer invoices and tried to match them up with pounds sold. But that was a tedious process and only as accurate as the invoices submitted by packers.

Shortly after being hired as assistant retail director, Eichar's sole job became tracking pounds sold. One day, he noticed a packer's shipments to a distributor were quite low. So, Eichar called the distributor to inquire about sales. To his surprise, the fresh-out-of-college Eichar learned the distributor was enjoying record sales of Certified Angus Beef. To verify that, the licensee faxed Eichar fifteen invoices from the packer.

"We matched the invoices up with our records and we were missing about ten of the fifteen invoices they'd faxed over," Eichar says.

After conducting an investigation into the missing invoices, Eichar learned the ten had all come from one packer, which was contractually obligated to submit those invoices to Certified Angus Beef. Colvin called the packer, who was an upstanding licensee, and told him about the missing invoices. The packer, baffled by the inquiry, just asked, "What missing invoices?" He then looked through the company's records and not only found the ten invoices missing during a five-month period in 1988, but discovered sixty more that had not been reported to Certified Angus Beef. That amounted to $15,000 in unpaid commissions to the not-for-profit company.

Certified Angus Beef distributors take pride in the brand and take every opportunity to educate their staff and customers. At George Wells Meat Company, licensed in 1983, Certified Angus Beef officials watch and learn as their partners teach. From left to right: Dr. Bob VanStavern, Dr. Erskine Cash, Mick Colvin, county extension agent Chet Hughes, and a George Wells representative.

1987

Retail sales exceed foodservice sales, proving quality important for at-home meals

At annual conference, Rick Nicholas, Newport Meat Company, encourages foodservice distributors to market chuck and round cuts, which are only pennies more than USDA Choice; result is greater carcass utilization

USDA Colorado Direct Cattle Close identifies premiums for qualifying cattle

First close-trimmed products: Champion Boxed Beef (formerly Litvak), strip loins

Eichar says nobody accused the packer of wrongdoing or of intentionally not reporting invoices; it was thought to be an administrative error. But the fact remained, there needed to be a better tracking system to catch situations like this one. Thus, the monitoring division, later renamed trademark compliance, and finally brand assurance—the policing arm of Certified Angus Beef—was formed. And Eichar, who had a passion for numbers and computers, was charged with overseeing the division's development along with a computer program that could track every pound.

"When I arrived, there were two computers," Eichar says. "We had not started using word processing software, so I was the first one to help facilitate the installation of a word processor. Up until that time, they used carbon paper and typewriters."

And now, the company was transitioning from yesterday's communications methods and implementing a state-of-the-art tracking system. Eichar interviewed computer dealers and the consensus was that it would take $20,000 to develop such a computer network. Eichar knew that was too much money for Certified Angus Beef, but estimated he could do it on his own for just under $5,000, despite not having any computer networking or programming experience. Colvin trusted Eichar, knowing the twenty-three-year-old could do anything he put his mind to.

Eichar had taken one computer-programming class in college and began reading books about computers, software, and programming. When he was ready, he modified already written code to meet the company's needs and installed the first network for less than $5,000. The program he wrote broke the carcass up into sections; for example,

Always an educator, Dr. Bob VanStavern frequently speaks to chefs and retailers about the brand and the importance of marbling. His "Science Behind the Sizzle" presentation becomes the basis for many customer-training programs and an approved American Culinary Federation seminar.

Science Behind the Sizzle

At the company's foundation are its scientific standards. To convey that, Dr. Bob VanStavern wrote "Science Behind the Sizzle" in 1988 and presented it at the brand's annual conference to explain the merits of Certified Angus Beef products. It discussed the traits of Angus cattle and the science behind marbling (including when and how intramuscular fat develops), the chronological age of cattle, and the relationship between fat and cholesterol.

"We had to get into some of the science that we had based the standards on, to get people to understand that this wasn't just a run-of-the-mill deal. This wasn't smoke and mirrors. This wasn't somebody coming up with a fancy name to sell Angus cattle. This thing had value," VanStavern says.

He gave presentations at various culinary functions and in 1993, he presented at the World Angus Forum in Rapid City, South Dakota. By 1996, "Science Behind the Sizzle" had received American Culinary Federation accreditation for continuing education hours. It became the standard presentation for staff to train thousands of chefs and foodservice sales representatives.

the program accounted for it having two chucks and two rounds, and approximated weights for each cut. A traditional programmer, who did not have thorough meat knowledge, would understandably miss such nuances.

For the next twelve years, trademark compliance employees manually entered data from paper invoices into the program. In 2000, company programmers made it Windows compatible and named it TIPS, or the "Trademark Integrity Protection System." Now, distributors send Certified Angus Beef data files that automatically upload into the TIPS database.

Since its creation in 1988, product monitoring has helped Certified Angus Beef collect $1 million on unpaid invoices for fifty million pounds of product. Equally important, the program supports the brand's claim that every pound is tracked from certification to the consumer. Some believe that without the development of this tracking system, Certified Angus Beef would not be the driving force it has become. The sheer size and complex business model of the company make such a program an invaluable tool to the company's success.

"TIPS really built the checks-and-balances to audit packers, audit distributors, and ultimately audit the end-user restaurants and retail stores," Eichar says.

While TIPS was an incredible advent for the company, it was still quite limited. TIPS users could only view the product data. The program did not audit the use of the Certified Angus Beef logo. But that was where a vigilant staff came in handy. They challenged anybody who misused the logo or the name. One such case involved a Chicago-based meat purveyor who was marketing "Certified 100% Angus Beef" but was neither a licensee nor had proof of such claims.

Tapping into America's strong affection for mascots, Certified Angus Beef introduces "Andy Angus" in 1987. He appears at food shows and product debuts in grocery stores, where he meets up with other mascots.

"We believe the words 'Certified Angus Beef' cannot constitute a trademark in themselves, since they perfectly describe the goods," wrote Kevin E. Snapp, a lawyer for Gourmet Bazaar, in a 1986 letter to the original Angus brand's lawyers.

Technically, Snapp had a valid point. The words and the logo of Certified Angus Beef were not trademarked. The logo itself was a certification mark filed August 3, 1979, and granted principal registration April 19, 1983. According to the U.S. Patent and Trademark Office, the mark certified that meat from Angus cattle met Certified Angus Beef's requirements and was approved by USDA. But the differences between a certification mark and a trademark were so minuscule that either side could be right.

By the government's definitions, a certification mark is used with the owner's permission by someone other than the owner to "certify" goods or services to meet a defined standard, while a trademark distinguishes goods or services of one manufacturer from the goods and services of another.

The "Andy Angus" mascot becomes a popular marketing symbol. It is used on grocery bags, like this one, and in advertising.

The difference between a trademark and a certification mark would prove to be the company's biggest challenge in its second decade. A North Carolina employee from Houston-based Sysco Corporation, the world's largest foodservice distributor and a multibillion-dollar company, used the Certified Angus Beef logo in a side-by-side comparison with another beef brand on a sales sheet. The not-for-profit company sued Sysco in October 1992 for misusing the company's logo. Sysco countersued Certified Angus Beef claiming the logo was a certification mark. While America was embracing its new president, Bill Clinton, Certified Angus Beef was taking on the Goliath of the foodservice world. The verdict of the case would either catapult the company into the future, giving it solid precedent to protect the brand, or it would possibly cripple the American Angus Association's successful subsidiary.

During the 1980s, the staff teams up with distribution partners to promote the brand at the National Restaurant Association and Club Managers Association of America shows. Frequent participants include: Jeff Stewart (top, second from left), Sharon Foster (fourth from left), James Sisk, Nancy Matheson-Burns, Arnie Hein, and Bill Wylie. Below: Staff member Bill Wylie discusses the brand with guests at a food show booth.

1988

Monthly sales record in January with five million pounds

Supply Development Division begins

First "Big Three" packer: Excel Corporation (Cargill Meat Solutions)

Dr. Bob VanStavern presents "Science Behind the Sizzle" at annual conference

One-millionth carcass certified to date

Sales exceed 50 million pounds during year, with products available in 18 countries

First international restaurant chain: New Tokyo, Japan, with 38 locations

Licensed restaurants and retailers all over the world sport the Certified Angus Beef logo. Research shows consumers positively react to the brand identification, boosting licensee product sales.

5 Protecting the Brand

In 1992, Certified Angus Beef proved it would protect its brand by any means necessary. On October 15 of that year, the American Angus Association filed suit against Bell Sysco Food Services Inc. and Sysco Corporation for trademark infringement. In a common legal tactic or maneuver, Sysco filed a countersuit shortly thereafter.

Brent Eichar was standing by the fax machine when the countersuit arrived. Back then, fax machines—at least this fax machine—used paper rolls instead of sheets of paper. As he watched the first few feet print out and realized its content, he became slightly nervous, maybe even scared.

"We were a small company with maybe twenty employees at the time," Eichar says. "If we lose this thing, Certified Angus Beef quite possibly goes away."

The reason for the Association's complaint began on September 1, 1992, when Sysco marketed its new Supreme Angus Beef program with an obvious intent to compete with Certified Angus Beef. The company created a color sales brochure that was given to restaurants. It said:

Only the best cuts wear the Supreme Angus Beef Brand. Our Program is based on the toughest specifications in the industry in which the beef undergoes THREE inspections, not just two. ... These inspections are conducted by not only USDA, but special SYSCO Quality Assurance inspectors specifically trained to work Angus Beef. Our inspectors hand-pick yield grade three, A-maturity or better cattle, so only the youngest (no more than 30 months) become SYSCO Supreme Angus Beef. Then it's graded Choice or Prime and guaranteed Angus Beef by the USDA.

1989

First quarter-inch-trim product: Kenosha Beef International

Frankfurter available from Independent Meat Company

First on-farm national training program: Boot Camp (later called Roundup)

National Supermarkets licenses 58 stores for all steaks and roasts to become largest retailer; stores later acquired by Schnuck Markets

The Art of Cooking with Certified Angus Beef cookbook features recipes from 36 world-class chefs

New Carcass Data Collection Service evaluates more than 3,000 cattle during year

First fully cooked product: Hudson Meat Company, deli top round

According to federal court records, USDA wrote Sysco on September 24, 1992, saying, "USDA does not, nor can it, guarantee that the beef being sold is 'Angus Beef.' Please take steps to correct your advertising by removing the statement 'Guaranteed Angus Beef by the USDA.' " Evidence showed a Sysco salesman used the term Certified Angus Beef in conjunction with Sysco product, which increased confusion among restaurants, the Association argued. Furthermore, in a Supreme Angus Beef Price List, which had been distributed throughout North Carolina, Bell Sysco had copied the Certified Angus Beef logo for a comparison, but neglected to use the word "versus." Nonetheless, Chief Judge Richard Voorhees, U.S. District Court for the Western District of North Carolina, wrote Sysco's litigation was *not* that the company "cannot obtain Angus beef, but that they want to use the phrase 'Certified Angus Beef' to describe and promote their Angus beef program."

On the first day of litigation, Voorhees granted a temporary restraining order against Sysco. That barred the company from further unauthorized use of the logo, from Supreme Angus Beef comparisons to Certified Angus Beef, and from falsely representing any affiliation with Certified Angus Beef. The order also banned Sysco from inaccurately saying Supreme Angus Beef was guaranteed Angus beef by USDA. The judge later said Sysco was not entitled to use the words "Certified Angus Beef," even without the logo.

"We thought we caught them with their hand in the cookie jar," says Richard Rossier, partner at the McLeod, Watkinson & Miller law firm. "They were copying our marks and putting it on their documents in their introduction of Supreme Angus Beef."

But Sysco argued in its counterclaim that the term, "Certified Angus Beef," was not trademarked; rather, it was a certification mark. Sysco sought to cancel the certification mark on the grounds that it does not describe a business format, but rather represents the certification of certain end products. The court dismissed that argument.

"It was a certification mark, but it was still a mark and we had legal rights in the mark," Rossier says. "Ultimately, we were ready to go to trial."

The case would never go to trial. Although the two parties settled out of court for undisclosed terms, the judge ruled in favor of most of Certified Angus Beef's motions. The case lasted more than two years and Colvin won yet another crucial challenge. He called it the biggest and most important victory of his Certified Angus Beef career. Eichar says Colvin's quiet confidence instilled a lot of peace within the staff. There was never a panic, he says.

"We knew we had done the right things," Eichar says.

As for Sysco, the company is now Certified Angus Beef's largest foodservice distributor. From the late 1990s to the mid-2000s, it purchased more than a dozen meat purveyors throughout the country, many of which were Certified Angus Beef licensees, including Buckhead Beef Company, Newport Meats, Malcolm Meats, and Freedman Meats. The two companies work well together, officials say, and there are no hard feelings from the litigation. "Sysco is very proud of Certified Angus Beef," says Tim Hussman, president of Newport Meats.

Andy Malcolm agrees. Since Sysco purchased Malcolm Meats in 1999, he says, the partnership with Certified Angus Beef and Sysco has been a "perfect opportunity." He says both companies' top officials meet annually at Sysco corporate.

"We meet on a regular basis to talk about our strategies and understand more and more about Certified Angus Beef," says Malcolm, vice president of Sysco corporate and chairman of its Specialty Meat Companies. "It's a very compelling relationship."

As for the lawsuit, he says, "People get over those things."

Just as Disney protects its trademarked mouse ears, Certified Angus Beef does not tolerate misuse of its logos. The company proactively works with licensed partners to make sure customers understand how to use the logos and rectifies prohibited uses by unaffiliated businesses.

Brand Assurance

Amanda Barstow has been called everything from the beef cop to the gestapo. In her role as brand assurance director, people have actually challenged her to fights. But little really gets to her. It's her job to make sure people are not stealing the Certified Angus Beef name and that licensees are properly using it. In addition, Barstow manages the company's program of registering trademarks throughout the world. While those tasks may sound mundane, they can actually be quite exciting.

Once, Barstow was tipped about a Texas restaurant using the Certified Angus Beef logo on the front of the building. The logo was identical except the steer was turned to face the opposite direction. Barstow politely called the restaurant owner, who balked at the allegations saying the sign was nothing like the Certified Angus Beef logo. And then he went on a tirade. According to Barstow, the man said: "Who do you think you are? You're just a big corporation picking on a small guy."

The man said he never went to college—"I'm not very book smart, but I'm street-smart. And if you want to come down here, we can just fight it out."

The seriousness sank in.

"When you're on the phone and they're yelling and cussing at you, our goal is not to make the customer mad; that's the last thing we want to do. We still consider them a customer until we have to take legal action," Barstow says.

Nonetheless, she could not contain the Texan. Rossier would call him and receive the same treatment. Finally, it was concluded litigation was the only course. Rossier contacted a lawyer with trademark experience in the restaurant's town to see if he could assist with filing a temporary restraining order. As it turned out, the restaurateur was a client of the local attorney, who asked, "What did he do this time?"

Upon hearing the story, the lawyer talked to his client and quoted trademark law. The man who had wanted a boxing match with Barstow took his sign down.

That's an extreme case file from the brand assurance division. Most of the time, when a restaurant is improperly using the Certified Angus Beef name or logo, Barstow explains the program, that it is a federal trademark, and that they can't use it unless they're licensed. She and her team take the opportunity to educate them about the brand, its scientific specifications, and the meaning of USDA certified. The restaurant owner may say they are buying the product, in which case Barstow checks TIPS, the Trademark Integrity Protection System, because every pound of Certified Angus Beef is tracked. If they are not in the system, she checks with area distributors. If they're not a licensed distributor's customer, she finds out where they buy. Then, she writes to the restaurateur or the distributor that the brand name is being misused.

Trademark Territory

Although USDA approved the brand's specifications in 1979, trademark approval for this certification process took a bit more time. The logo was registered by the U.S. Patent & Trademark Office in 1983. The same mark earned international recognition in Canada in 1986. The first Asian approval came in Taiwan in 1990, and while the petition had been in place for many years, the logo received the okay in Japan a few years later.

Also in the 1990s, as the trademark lawsuit with Sysco was unfolding, a refined "shield" logo and banner mark began to be used for product promotions beyond USDA certification procedures. The shield logo finally gained U.S. registration in 2002, just as the premium logo began to appear on point-of-sale materials in retail stores and restaurants. It achieved U.S. registration soon after.

Additional trademarks include slogans such as "Angus Beef at its Best" and "Clearly a Matter of Taste." Canada was the first country to approve the three-word "Certified Angus Beef" trade name in 1995.

First Logo, 1978

Certification Mark, 1983

Shield, 1992

Brand Mark, 2001

Banner, 1995

Japan Mark, 2003

Prime, 2002

Natural, 2004

Prime Natural, 2009

Logo placement and the product description are critical on menus. Most restaurants design their own menus, but occasionally they request help from Certified Angus Beef's design team.

1980s menu with certification mark

1998 menu with shield logo; first menu staff designed.

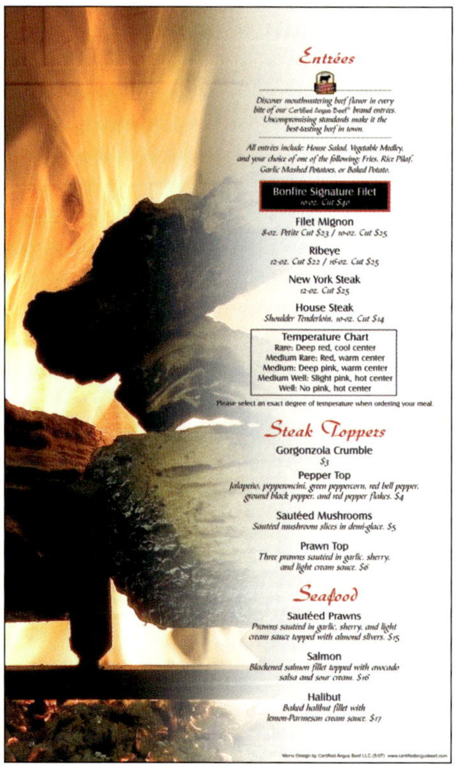

Many menus, such as The Bonfire Grill's, offer steak selection and doneness guides.

2009 menu for Waldorf=Astoria in Orlando; first to identify all four product lines.

Occasionally, the restaurateur tells Barstow she cannot dictate what he puts on the menu. "We explain, 'I understand how you might feel, but you can't put Certified Angus Beef on your menu without being licensed.' " The really hostile cases reject the thought of taking the name and logo off the menu, to which Barstow replies: "I would be like you if somebody out of the blue called. But here's our name, here's our trademark registration for that mark that you're using. Look up a trademark attorney and call them."

Most of their issues—91 percent—have to do with restaurants misusing the mark. That's why brand assurance conducts regular restaurant and retail audits. Legal action is the last recourse.

"Sometimes restaurants are buying the product and they don't know they need a license to use the logo, so we just have to educate them and get them signed up," she says. "Other times they know darn well that they're not getting our product."

She says brand assurance staff spends most of its time sorting data in TIPS, looking for anomalies in sales numbers, or dramatic dips or spikes in middle meats or grinds.

"Just as important as the numbers are the people," Barstow says. "Long-time staff members like Sharon Swanson play a pivotal role in customer relations. Most of our team has been here longer than ten years, yet they still look to Sharon for her twenty-plus years of leadership."

On the retail side, the process has been much different. The staff tracks point-of-sale materials and product offerings by comparing what was in stock or sold versus TIPS purchase records. In addition, because retail selling is so much about presentation, store visits make sure the displays accurately portray the brand.

"Historically, if the case didn't look good, we'd address that in a letter to the corporate office and have them take corrective action, try to get that store location on board; and if not, then we canceled that location," Barstow says.

As retail brand assurance evolved, it changed enforcement philosophies along with the company's retail licensing rules, as noted in Chapter 7. "We have learned much from our licensees, and continue to do so," president John Stika comments. Current licensing standards allow more flexibility, but the auditing team still makes sure the grocery stores are meeting volume requirements.

When Certified Angus Beef catches misuse of the logo, it raises the issue. The regional account managers often intervene to further explain the brand's official message. But in the end, Barstow says, all licensees are appreciative of the brand assurance efforts because they know, "We are going to maintain the integrity of the brand at any expense."

However, they want to keep everybody happy, too. When Barstow and colleague Marylynn Roe analyzed the most frequent issues, it seemed many could have been prevented if there was an ear for licensees. The account managers' job was to get new accounts and train existing ones. Brand assurance operatives were there to protect the brand. Whose job was it to help with the little problems?

In 2005, the company established a customer service team headed by Roe. From then on, customer service would send out license agreements and take point-of-sale orders. Customer service representatives would also handle general queries and serve as

Through her dedication to Certified Angus Beef and its licensees, Marylynn Roe works with Barstow to create the customer service team, forming a vital link between the company's regional staff and licensed restaurants and retailers.

Certified Angus Beef has long advertised in grocer and restaurant trade publications. Advertisements prominently display the brand's logo and illustrate the great-tasting beef. In 2007, the ads came full circle, connecting the brand with Angus producers.

2003 Foodservice and Retail

1980s Foodservice

1980s Foodservice

2007 Foodservice, with Chef John Doherty and purebred Angus producer Kevin Yon

2007 Retail, with Ed Steinmetz from Giant Eagle, purebred Angus producer Kevin Yon, cattle feeder Mark Sebranek, commercial beef producer Joe Davis, and brand president John Stika

From the beginning, promotional materials have set the brand apart from other meats. They help retailers demonstrate their passion for providing customers with top quality.

Above:
1980s in-store signage

Opposite, top:
1980s grocery bag

Opposite, bottom:
1992 case display with "Clearly a Matter of Taste" slogan

liaisons for the regional account managers in the field. Most importantly, there would be a central point for licensees.

"Marylynn was saying we need customer service, so we can control these point-of-sale orders," Barstow says. "We can get to know these customers better and know what they need."

Trademarks and Registrations

Registering the brand name and logo in other countries may be the biggest challenge for brand assurance, Barstow says. Even the United States requires a lot of paperwork and trademark attorney time. In foreign countries, especially, they don't understand "Angus," Barstow says.

"They don't understand brands of beef. Some countries will register the logo, but not the name," she says. "When we register a design, we want to also protect the words 'Certified Angus Beef' in the market, so if a country proposes to drop any part of the three-word phrase from the registration, we'll say no, we won't do that."

Canada recorded the company's first international logo registration in 1986. In the late 2000s, more than seventy countries have at least one Certified Angus Beef registration. While the brand's logo was federally registered in the U.S. in the 1980s, the redesigned version received registration in January 2003. The updated logo enhanced the colors and included "since 1978" at the top and "brand" at the bottom. The words "Certified Angus Beef" were inside the logo's waving banner. The modern logo is registered for use with marketing services promoting the sale of qualified beef products for meat packagers, distributors, wholesalers, and retailers.

In 2009, there were ten pending registrations, including slogans, worldwide. Barstow says Hong Kong was the toughest campaign of them all, but the Chinese state capitulated in 2007.

Certified Angus Beef licensees can sell product internationally even where the trademark is not registered, Barstow says. But there's a risk of somebody using the logo without permission or creating a similar logo.

"Some countries are based on first use of the mark in the market and others are based on first registration in the market," she says. "In these cases, we can prove if we've used our marks in the market, meaning we sold product first into that market, but if the other company beats us to registration, they may still get it and own the registration."

Some international restaurants and distributors have simply copied and pasted the Certified Angus Beef logo on their menus and sales catalogs. Fortunately, the brand was tipped as they tried to register the mark.

"We were able to talk to the owners and tell them we own this mark and that they can't register it in that country because we own it already," Barstow says. "We had them legally reassign the mark to us and they became licensed Certified Angus Beef restaurants."

Angus Confusion

Over the years, protecting the brand has also meant facing other Angus programs. Several others have developed and created "Angus confusion" in the marketplace. But some like the competition.

"Imitation is the sincerest form of flattery," wrote Charles Caleb Colton, a famous British writer and cleric in the 1800s. Are Colton's musings true for Certified Angus Beef?

Today, it's nearly impossible to walk into a grocery store, fast-food restaurant, or random eatery and not see some kind of Angus burger or steak. In addition, various pet food companies claim to use Angus.

In the packinghouses alone, as of September 2009, there were sixty-two Angus brands. USDA's Larry Meadows says these programs create confusion for the grader. But in the eyes of licensees, they create confusion for the consumer.

For Doug Rodda, sales manager for DeBragga and Spitler, there's an emotional connection to Certified Angus Beef. "It's been my livelihood for more than twenty-five years." So when he sees a new Angus program popping up on the street, he goes to his office, makes a phone call, and sets up appointments with clients who have not been licensed very long. He wants to make sure they know what this program is about.

"There are so many Angus programs out there now that we constantly need reeducation on what makes this product different," Rodda says. "Every week and month that goes by there's a new product out there: Premium Angus, this Angus, that Angus. The performance of Certified Angus Beef is much better. New York chefs appreciate that."

Malcolm, of Sysco, believes Certified Angus Beef is a heavily recognized brand in the industry. "I think that is huge," he says. "There have been a lot of people who've tried to duplicate it, but those brands just never stick. This is the first time anyone has ever taken a fresh beef product, branded it, and created a value-added brand in commodity beef."

From his perspective, the product quality and the company's integrity override any "me-too" Angus brand. Certified Angus Beef's reputation is "everything to me," Malcolm says. "I wake up every day, come to work and oversee sixteen manufacturing facilities, and the leadership in those facilities reports to me. I know every day that my plants are producing a high-quality, high-integrity product. Money is nice, but you can't buy reputation."

For Certified Angus Beef, that reputation starts with being owned by the Association. While competing brands may be willing to cut quality to improve the bottom line, the Association wants to improve bull sales. Malcolm sees that as a huge advantage over the competition. "Owned by the Association has a clear alignment with all the Angus breeders," he says.

In addition to the integrity that communicates in the beef product trade, Buckhead Beef president Chad Stine says Certified Angus Beef's connections give Angus breeders a leg up on the rest of the purebred and commercial cattle industry.

"The only breed organization that has the branded beef program with enough supply and volume to be able to share information back upstream all the way to the production level is Certified Angus Beef," Stine says. "That transfer of information has been vital to the success of the American Angus Association."

On all Certified Angus Beef retail products, the traditional, Natural, and Prime lines are clearly identified. The company provides retailers with specific marketing materials for each category.

Opposite: Certified Angus Beef licensees appreciate the product's quality and association with Angus producers. They tell customers other Angus brands are not the same. Food City, owned by K-VA-T Food Stores, and Montana Mike's urge customers to avoid imposters.

However, some breeders favor other, privately-owned and for-profit Angus brands. Their argument is that anything "Angus" is good for the breed.

It's easy to find dissenters to that view, and they say the other brands are somewhere between irrelevant and useful. Certified Angus Beef cannot control quality specs for other brands, and Angus, by itself, is a common word that cannot be trademarked. Those who ride for a while on the original brand's coattails will sell some products, but if they fall short in one area or another, it's hard to build much of a following. Still, competition is healthy.

In the retail markets, the other Angus programs are more of a threat. Nearly every major grocery chain has its own brand. But Bob Boehringer of Tom Thumb says they don't compare. "They're all Select beef with a logo," he says. "Look at a strip steak and there's not a speck of fat [marbling] in it except on the border."

As for the restaurant industry, some restaurants won't put the Certified Angus Beef logo on the menu because they worry about confusion with the more generic "Angus" use on fast-food menus. Others have the confidence of experience, and their customers appreciate seeing that Certified Angus Beef icon of excellence. Take Paul Mackay, owner of El Gaucho restaurant, licensed since 1996, in Seattle. His daughter works for a major distributor and sells another Angus program. For years, she's tried to convert him. He always says no.

"I dance with the one that brought me to the party," Mackay says.

1990

Licensed restaurants exceed 1,000

First television commercial: National Supermarkets

Licensed packers access 30 percent of nation's fed cattle

Two million Angus-influenced cattle identified during year

Trademark protection expanded with registrations in Puerto Rico, Mexico, and South Korea

Feedlot Gain and Carcass Contest (Value Discovery Program) begins at Decatur County Feed Yard

Shabu-shabu *is a Japanese beef recipe that uses thinly sliced red meat, usually chuck roll. Certified Angus Beef's abundant marbling makes it a preferred American beef brand for this dish in Japan.*

6 INTERNATIONAL IN-ROADS

In 1994, when Colvin spoke at the Angus Australia National Show & Sale, he helped producers "Down Under" see that they were not just in the cattle business, but in the food business as well. Colvin waxed eloquent about how the Certified Angus Beef program worked, its science-based specifications, and its market influence in sending premiums all the way to the producer.

For many attendees, the presentation was just a Yankee blabbing about American Angus quality. But for Michael Pointer, a relatively new cattleman and veteran sheep producer, Colvin's speech resonated.

"Mick blew me away," Pointer says. "I thought, why on earth haven't we been doing something similar in this country?"

Immediately after Colvin's speech, Pointer studied how Australia could imitate America's Certified Angus Beef. Much like the Colvin experience, many colleagues told him such a program would not work. And, like Colvin, that only encouraged him more. Then, two years later, the Australian Angus Society received a letter from a restaurant operator wanting to know where he could purchase Angus beef. That was Pointer's chance. He had to capitalize on an "out of the blue" coincidence. The chef loved the meat and has used it since.

The Australian Angus Society created its own brand and went from processing twelve animals in 1996 to more than a hundred thousand per year. That's about 4 percent the volume of the original Angus brand in America, but the Australian industry is much smaller.

Located in Taiwan, Chu Shao Guo Wu Liai Li's seven restaurants use boneless short ribs for sukiyaki. *Outdoor signage, posters, and placemats promote Certified Angus Beef, helping it earn the brand's international marketing award in 2008.*

Just like Certified Angus Beef did in the States, the Australian Angus-focused program greatly increased demand for Angus beef and the value of bulls. Also like the Americans—when their program became USDA's first certified one—Pointer's group influenced the country's agricultural policy by creating a need for a national grading system.

"Virtually every Angus society in the world is setting out to run an Angus branded beef program, and I think that is due entirely to Certified Angus Beef," Pointer says. "It has changed the face of the world beef industry."

However, it would be difficult for another country's Angus society to set up a similar program because the original and primary Certified Angus Beef brand's trademark is either approved or pending in most countries. Only Canada, Japan, and Australia have comparable grading systems to the United States. Nonetheless, the international markets make up nearly 10 percent of Certified Angus Beef's overall sales and there is great potential for continued growth.

As noted in Chapter 3, those international roots started with The Bruss Company and DeBragga and Spitler in the early 1980s, when they were selling product to their customers in Singapore and Bermuda, respectively. Their early successes led to interest from export-focused distributors. The first export license was granted in June 1985 to Midamar Corporation, followed by Oritz International in January 1986, and Day-Lee Meats in January 1987.

By 1994, Certified Angus Beef was in twenty-one markets, including the Bahamas, Bermuda, Canada, China, Mexico, the Caribbean Islands, Japan, Taiwan, Bahrain, Moscow, the Philippines, Indonesia, Singapore, South Korea, Hong Kong, Thailand, Peru, and the British West Indies. Fifteen years later, more than three thousand licensees were offering products in many more countries.

Businesswise, international markets bring notoriety to the Certified Angus Beef brand. There's a great marketing advantage when a U.S. brand can be promoted as sold all over the world. Besides, export sales are comprised of different cuts than those most in demand in the U.S. And in winter's slower beef months, vacationers in tropical locales continue to drive demand for top-quality steaks. Those factors help balance carcass utilization, creating greater value for packers and producers.

Certified Angus Beef establishments include such famed locations as the Sandy Lane Hotel in Barbados, where Tiger Woods was married in 2004; and Rare, about which *Frommers.com* says, "We'll go out on a limb here, and claim this Dino Jagtiani steakhouse is the best and most classic 'chophouse' in the Caribbean. ... The chefs here are the only ones in the Caribbean to offer Certified Angus [Prime] Beef."

And then there's the internationally renowned grocery chain, Loblaws, which markets the Certified Angus Beef brand in more than 220 of its Loblaws, Fortinos Supermarket, and Zehrs stores in Canada's eastern provinces. But Colvin says he didn't originally consider selling outside of the U.S. And for good reason. In the early 1980s, he was just trying to keep the program above water.

"I was more worried about foodservice and retail," he recalls, "but after talking to Butch Pfaelzer of The Bruss Company, I realized the potential."

With that said, Colvin tends to understate his role in the development of the program's international status and everything else, says Tracey Erickson, vice president of marketing and former export director.

"Mick never gives himself enough credit, which is part of why we all love him so much. He's the ultimate example of a humble servant leader," Erickson says. "He got international off the ground."

After Colvin, Homero Recio, and Alan Waggoner worked the international scene into their other duties. Then Erickson, fresh out of Colorado State University, became the first full-time export director in 1992. Her thesis on the Asian marketplace and its multiple distribution layers proved useful right away. Japan had liberalized trade in 1991, making it a prime market for high-quality U.S. beef. License agreements and distribution policies were needed to ensure the brand's identity in the market.

Once those agreements were in place, the international story is really no different than the rest of the company's history. Licensees were won over by the quality, integrity, and potential success of the program. And once they took on the brand, they carried it to great heights. For Eliott Reyes Sr., who initiated the communication through a letter to Colvin in 1986, Certified Angus Beef has absolutely redefined his foodservice and retail distribution company Provimi de Puerto Rico.

"Certified Angus Beef is the bread and butter of our products," Reyes says.

1991

Ten million Angus-influenced cattle identified to date

Packing companies compete to increase carcass utilization; awards recognize BeefAmerica-Omaha, Beef Specialists of Iowa, Independent Meat Company, and Excel-Sterling

First distributor in Canada: Macgregors Meat & Seafood. J.J. Derma Meats (later a Sysco affiliate) licenses next spring

First distributor in Bermuda: Butterfield & Vallis

USDA transfers program monitoring from Food Safety Inspection Service to Agricultural Marketing Service

Asian Invasion

Early successes in Asia were driven by licensed exporters, like Bruss and Gerber Agri-Export. After the Shangri-La in Singapore in 1983, Bruss supplied products to

Photo credit: Fred Minnick

Export Details

MacDonald Meat Company in Seattle, a licensed exporter since the mid-1980s, will "cut to order" steaks for its international clientele. As for what they want and when, it depends on the customer. But typically they're buying steaks and Asian favorites like short rib, plate, and chuck roll. Those diverse needs are reflective of the international division's important role to help merchandise some of the lesser-used cuts in the U.S.

Timing, delivery method, order size, and documentation also make export a bit different from U.S. distribution. A local restaurant may need one hundred pounds of steak for same-day delivery, whereas Kris Black, the chief operating officer for MacDonald, says the typical export order is about fifty thousand pounds and takes a full month to fulfill.

She and her team assemble the order and arrange shipment of the frozen cuts. "If it's a sea shipment, we'll load the containers here. They'll pick up the container, load it, and deliver it to the port," she says. "If it's an air shipment, we take it to a freight carrier to fly it overseas."

MacDonald's products are primarily destined for the Philippines, Hong Kong, and Vietnam. Every country has its own beef export verification program with USDA, and it is audited twice a year. It is a voluntary program, "but if you don't have a program, you cannot export to certain countries."

thirty-eight New Tokyo Restaurants in Japan. Gerber licensed the first of twenty Victoria Station restaurants in Tokyo in 1991, just as the market opened.

"As a company, we didn't try to sell beef in Japan in the 1980s," Erickson says, "Government purchase orders made it possible, but market liberalization really drove our opportunities for growth."

The first step was to cut through the multiple distribution layers in Japan to work more directly with restaurants and grocery stores. This was critical to ensure product quality for consumers, and licensed exporters with market know-how led the way. Under the direction of Earl Skahill, Gerber held training sessions for chefs in Japan, Korea, and at the newly licensed Grand Hotel in Taiwan in 1992.

Certified Angus Beef cuts are sampled in a Loblaws store. When Canada's largest grocer begins carrying the product in 1994, it helps Certified Angus Beef carve out a new market.

In these markets, traditional dishes used beef in *shabu-shabu* and *sukiyaki*. In Japan, for example, the average consumer ate less than ten pounds of beef each year, but he or she required highly marbled cuts, like short ribs and chuck rolls. They favored Kobe beef, from Japanese Wagyu cattle raised in the Kobe province, but Erickson says Certified Angus Beef never tried to compete with Kobe.

"They call Kobe beef, *shimofuri*, meaning 'sprinkled with snow.' That product is off the scale in terms of quality," Erickson says. "Our position was the best American beef, unlike here domestically, where we'd say it's the best beef you can buy."

The Japanese liking for the brand quickly showed. Within a couple years of focusing on the market, Japan became the number one export country. Still, staff found that doing business in Japan posed challenges because their consumers were so different from Americans. According to *Marketing Japanese Style* (Quorum Books 1995), Japanese consumers already know so much about the products that the key creative issue is how the message is said, or "pictured."

As evidenced with Pokemon—the wacky trading card and video sensation in the 1990s—Japanese consumers related to cartoon characters, so Certified Angus Beef executed various storytelling tactics. One successful campaign was called Five Star American Beef, featuring glitzy stars. Additional campaigns included Angus cattle photos—a direct connection to the ranch that became more common in the U.S. a decade later. That campaign was displayed in two hundred newly licensed stores in Japan in 1997.

While those marketing tactics were effective, perhaps the biggest reason for success of the Certified Angus Beef brand is its literal translation in Japanese.

"When they see that black animal on the logo, they translate it to black-haired beef. In the Japanese culture that signifies quality," says Maggie O'Quinn, an international staff member since 1997 and current brand representative for countries south of the U.S. border. "In stores, you see promotions for black-haired hogs and black chickens. Anything black is associated with quality when it comes to proteins."

Another marketing advantage Certified Angus Beef had in Japan was IBP (Tyson), which was licensed in 1992.

"They had a large international team and offices all over the world, but the Asia office, with Takamichi Tawara as president, influenced our growth the most," Erickson says. "IBP really drove the presence of Certified Angus Beef in that market." The brand hired its own consultant in Asia in 1996, then opened a Tokyo office in 2003.

Supplied by The Bruss Company, New Tokyo is the first licensed international restaurant chain. This steak fare promotion is one of many in Japan, which became a prime market after the country liberalized trade in 1991.

That same year, Japan purchased nearly forty-seven million pounds of product. Then, a Washington state dairy cow tested positive for bovine spongiform encephalopathy (BSE) in December 2003. Although there was no connection to high-quality exported beef, almost overnight, the most successful international market for Certified Angus Beef disappeared.

Some Asian markets closed to U.S. beef temporarily. Others like South Korea and Japan took longer to reopen.

"Our staff didn't wait around for reopening," Erickson says, "We kept up communications with our customers. We focused on education to keep our brand relevant. We wanted to have our foot in the door when each market reopened."

Ironically, BSE has been associated with relatively old cows. Because of Certified Angus Beef's "A" maturity specification, cattle are typically less than thirty months of age, so the prospect of BSE is especially remote. As the closed borders continued, several licensees—feeling the lost revenue every day—requested the brand create a marketing program to point out the product's immunity to BSE, to which Certified Angus Beef said no.

"I took a very firm stand with our licensees," says Jim Riemann, former president of the brand, who replaced Colvin in 1999. "We were not going to separate ourselves from the rest of the beef industry. We were going to stand together and fight this thing and do the very best we can; some licensees were mad and didn't think that we were supporting them in the way we needed to."

Finally in July 2006, U.S. beef was allowed reentry to Japan, and in 2007, $230 million of U.S. beef had been shipped. While Certified Angus Beef welcomed the reopening, market access was limited to cattle verified as having no more than twenty months of age.

The brand licensed the Japanese distributor Kanematsu Corporation, and Don Company Ltd, a fifty-three unit restaurant chain in 2008. A year later, Japan

Dean Blankenship and Maurice Vaughan (left to right), from Beaver Street Fisheries, the parent company of Bahamas Food Services, show off several Certified Angus Beef cuts at a 1995 food show. Since 1993, the Bahamas resort market has grown appreciably as a destination for the brand.

Vacation Spots

If you plan to take your leisure on a vacation-oriented island, chances are you will come across a Certified Angus Beef licensed restaurant. As early as the mid-1980s, resorts and top restaurants outside of the U.S. were purchasing Certified Angus Beef products from either a U.S. or a local company.

Since 1993, Bahamas Food Services, the largest foodservice and retail distributor on the island, has been licensed with Certified Angus Beef. It focuses on those high-end resorts and restaurants. The company sells a good deal of middle meats, and every ground-beef patty sold is Certified Angus Beef. Considering those hamburger meals are going for $25, O'Quinn says, "That's phenomenal."

The brand can be found at these Caribbean and Latin American travel destinations, to mention a few:

- Coral Sands Resort, Harbour Island, Bahamas
- Graycliff Hotel & Restaurant, Nassau, Bahamas
- Grand Cayman Marriott Beach Resort, Cayman Islands
- Half Moon, Jamaica
- Hotel Quinta Real, Monterrey, Mexico and Guatemala City, Guatemala
- Hotel Veneto and Casino, Panama
- Las Mañanitas, Cuernavaca, Mexico
- Lyford Cay Club, Nassau, Bahamas
- Renaissance Aruba Resort & Casino, Aruba
- Sandy Lane Hotel, Barbados
- Westin Casuarina Resort, Cayman Islands
- Westin Dawn Beach Resort, St. Maarten

The Westin Casuarina Resort in Grand Cayman is renowned for beach barbecues featuring Certified Angus Beef brisket. From steaks in the upscale Casa Havana to beach-side dining at Tortugas, all three resort restaurants offer the brand.

Above left: Victoria Station is among the first licensed restaurants in Japan. It holds frequent brand promotions and expands to twenty locations by 1998.

Above right: Earl Skahill (right), from Gerber Agri-Export, works with Beef-America in 1992 to hold educational seminars in Korea and at the Grand Hotel, the first licensed Taiwanese account. Executive chef Urs Stauffacher (left), hosts the grand opening. Gerber plays a vital role in Certified Angus Beef's early success in Asia.

represented just 4 percent of the brand's international sales, a mere fraction of what it garnered ten years earlier. But brand executives believe Japan remains a viable and profitable market for the company in the future, especially if it reverts to thirty-month age verification.

While challenges loomed in Japan, South Korea reopened in June 2008 after U.S. officials agreed not to send meat from cattle older than thirty months. By the end of 2009, Korea was Certified Angus Beef's third-ranked importer with 9 percent of total international sales. The market reached pre-BSE levels in only five months.

Branded beef cuts are mainly found in stores in larger cities. Lotte Mart, which is the third-largest retail chain in the country, regularly promotes the brand in sixty-three stores. Its fall 2009 sale offered chuck-eye roll and boneless short rib. E-mart, Korea's largest retailer, also purchases products for its 119 stores.

"Our brand recognition and reputation have helped us penetrate the market," says Daisuke Shimojima, the brand's representative in Asia. "We are making positive connections in foodservice as well."

Asian licensees include the Marriott Resort & Spa in Guam and the Mitsui Japanese Restaurant chain in Taiwan, both of which offer the brand's Prime cuts.

Canadian Power

In the early 2000s, Lawrence Hawkins likened buying a steak at his abc Country Restaurants to purchasing an espresso when the machine was down or a bottle of wine from a server who couldn't open the bottle. In other words, his chain of Canadian family-style restaurants just didn't do steak well.

Above left: Pacific Overseas hosts a product viewing and tasting at Foodex 1993, an annual international food and beverage exhibition in Japan. Brand consultant Osamu Miyoshi and Butch Pfaelzer, from The Bruss Company, help Pacific Overseas host nearly 150 members of the Japanese food trade.

Above right: Yoshito Eto, of Foods Design Company Ltd., demonstrates cutting techniques at a January 1992 seminar in Tokyo. Hosted by the U.S. Meat Export Federation and Gerber Agri-Export, the seminar shares new merchandising methods for chuck clod and gooseneck with 144 guests. A Meat Export Federation member since 1987, Certified Angus Beef values the organization's leadership in U.S. beef education and promotion around the world.

"There was a time we didn't have a lot of confidence cooking steak," says Hawkins, who has sixty restaurants throughout Canada. "The comment cards would say: the steak was tough, attributes were not right, or too much gristle."

Then, in 2006, the franchise company licensed with Certified Angus Beef. Hawkins' franchisees were surprised to see they would not receive kickbacks as they do with other licensing agreements. But when they saw the increased sales and fewer returns, they were all about this new beef brand.

Specializing in comfort food in big portions, abc's menu is a brilliant display of wholesome ingredients at affordable prices. All beef products are Certified Angus Beef, including the cheddar bacon steakburger, Salisbury steak and mushrooms, pot roast dinner, and sirloin steak. Hawkins says steak sales were up 62 percent and steakburger sales jumped 54 percent the month after putting them on the menu.

"When people come to abc, they're not going to a steakhouse. A steak and prawn dinner for $20 is expensive for our segment," Hawkins says. "But if they have a good steak experience, chances are they will come back for lunch or breakfast. Now, with the Certified Angus Beef product, we get 9.9 out of 10. It is a tremendous value of steak. There are no more negative reviews."

The franchise chain certainly owes itself a good deal of credit for that success, training its waiters about the quality of Certified Angus Beef, mastering point-of-sale advertising, and maximizing use of the logo. Some success can also be attributed to the solid brand equity in Canada. In 2009, Canada was the largest international market for Certified Angus Beef, selling more than 26.8 million pounds a year or 38.8 percent of total international sales.

The first licensee was Macgregors Meat & Seafood in 1991, followed by J.J. Derma (Sysco Food Services of Canada/J.J. Derma Meats) the next spring. Some of the first restaurants included Bardi's Steak House and Black Angus Steak House, both in

Above left: Don Company Ltd. in Japan features the brand in fifty-three restaurants. Promotions include a spring 2009 shabu-shabu *fare. Sales in Asian markets mainly focus on chuck roll, chuck rib, and boneless short rib — cuts with limited market potential in U.S. restaurants and grocery stores.*

Above right: Export director Erickson (from left), IBP-Asia President Takamichi Tawara, and staff member Lori Franklin join efforts at a food-show booth. Tawara and IBP-Asia are invaluable allies to Certified Angus Beef after licensing in 1992.

Toronto. Restaurant chains also emerged over the years including Cactus Club Cafés, Joey restaurants, Milestones Grill & Bar, and Earl's. Among the chains, Shoeless Joe's menu offerings have grown from one Certified Angus Beef steak to several items with frequent promotions. Perhaps the most recognizable Canadian chain with a branded steak on the menu is Boston Pizza, with more than three hundred licensed restaurants nationwide.

In 1994, the country's largest chain of grocery stores, Loblaw Companies Limited of Brampton, Ontario, licensed its eastern stores. Longo's followed in 1997, with brand presence reaching the West Coast by 2005 at Country Grocer stores on Vancouver Island. Retail outlets also include M&M Meat Shop, with more than four hundred specialty food stores and a customer magazine that reaches one million households from coast to coast.

As those businesses continued signing on, Certified Angus Beef looked for ways it could capitalize on the Canadian cattle population for everyone's benefit. Unlike other countries the company was exporting to, Canada had a reliable foundation of Angus genetics and packing plants owned by U.S. companies. In 1999, the Certified Angus Beef Board voted in favor of investigating ways to work with the Canadian beef supply. However, the decision wasn't a slam-dunk. Internally, the board debated whether using Canadian cattle would devalue the U.S. herd. Steve Brooks from North Dakota, chairman in 1999, says some producers questioned how licensing Canadian packers would impact U.S. bull sales.

As Brooks and his colleagues researched the issue, they realized many U.S. cattle were being fed in Canada, and several Canadian Angus breeders were sending steers to the U.S. to be harvested. What really won the board over, he says, was the company's

With the brand in its sixty-three stores, Lotte Mart is South Korea's third-largest retailer and continues to expand. Store promotions include signage (inset) from Certified Angus Beef featuring black Angus cattle.

1996 annual conference, well before Canadian packers were licensed. "While in Toronto, we started seeing how much Certified Angus Beef was being sold in eastern Canada," Brooks says. "That was a big selling point."

The total herd was 10 percent the size of the U.S. herd, but about four hundred thousand slaughter cattle could meet Certified Angus Beef live specifications. If the brand could license Canadian plants, it could improve supply to a burgeoning sector.

"We were in a situation where we really needed more product," Riemann says. "We looked at the Canadian grading standards for beef carcasses, and realized they were so similar to what we have here in the U.S. that it would not be hard at all to go into Canada and certify carcasses as meeting the Certified Angus Beef specifications with just some minor tweaking."

Company representatives went to Alberta, Calgary, and met with the head of the Canadian Beef Grading Agency. Personnel there were excited about establishing a high-quality branded beef production process. In fact, just as it did in the U.S., Certified Angus Beef became the first to set up such grading criteria.

"We essentially became pioneers in starting up a beef carcass certification process there," Riemann says. "Canadian packers were very eager to work with us and were extremely cooperative."

In 2000, IBP-Lakeside in Alberta became the first licensed packer, followed by a Cargill plant at High River, Alberta. In 2009, there were four licensed Canadian plants, including the aforementioned Lakeside facility under the ownership of Calgary-based XL Foods, as well as St. Helen's Meat Packers in Toronto, and Cargill's Better Beef plant also in Ontario.

It turned out that Canadian packers helped the program during the BSE crisis. Despite having past BSE problems of their own, Canadians were nationalistic about

1992

National Beef Packing first packer to transition from packer roll to bagged subprimal identification

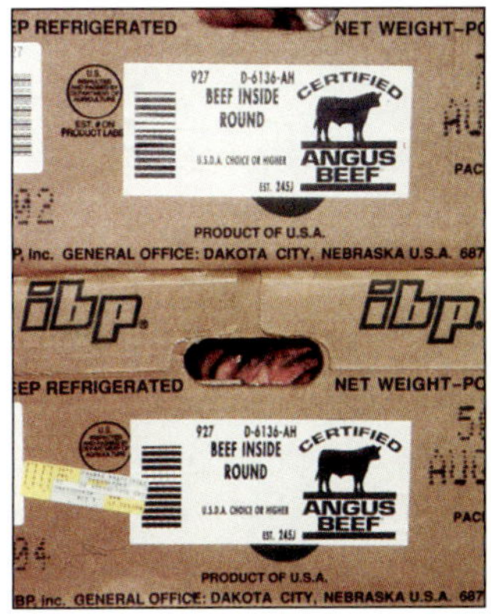

Production begins at IBP (later Tyson Fresh Meats)

New Export Division focuses on stronger international presence and balanced carcass utilization

Gerber Agri-Export and BeefAmerica hold educational seminars in Korea and at Grand Hotel in Taiwan

First producers recognized for dedication to beef quality: Summitcrest Farms, Bar S Ranch, and 37 Cattle Company

First distributors licensed in Mexico, Comercial NorteAmericana; and Taiwan, Shuh Sen Company Limited and Tzy Yao Enterprise Company Limited

country classics

certified angus beef® burgers

Served with never-ending wedge fries or a garden salad! Substitute a Caesar salad or onion rings

Cheddar Bacon Burger
With smoky bacon & Canadian cheddar

Classic Deluxe Burger
Like the name says!

M & M Burger
With mushrooms and melted mozzarella

Canadian Burger
Two Canadian beef patties, smoked ham & cheddar cheese

sandwiches & more burgers

Served with never-ending wedge fries or a garden salad! Substitute a Caesar salad or onion rings

Classic Country Clubhouse
Roast turkey, smoked bacon, lettuce & tomato

Hot Turkey Sandwich
Roast turkey and sage dressing layered on grilled French bread smothered in country gravy

Beef Dip
The taste difference with this all-time winner is tenderly roasted Certified Angus Beef®

Steak Sandwich
Our Certified Angus Beef® top sirloin comes with mushrooms on grilled garlic focaccia

Pulled Pork Sandwich
abc's famous pulled pork sandwich. Slow roasted pork so tender it pulls apart with a fork. On a sweet bun with sautéed onions and honey gold BBQ sauce

Chicken Club Burger
Charbroiled Fraser Valley chicken breast topped with smoked bacon

Mediterranean Chicken Burger
With spinach & feta cheese

Gardenburger
A patty made with mushrooms, vegetables, mozzarella and rice. Try it with BBQ sauce!

Healthy Choices. We cook our fries in zero trans fat Canadian Canola oil.

country heroes

Pot Roast Dinner
Certified Angus Beef® makes the difference. It's fall-apart tender

Salisbury Steak & Mushrooms
A lean, juicy charbroiled ground beef steak smothered in mushrooms and sautéed onions, with mashed potatoes & country gravy

Canadian Burger
Two Canadian beef patties, smoked ham & cheddar cheese

Beef Dip
The taste difference with this all-time winner is tenderly roasted Certified Angus Beef®

Tender, juicy and delicious Canadian Certified Angus Beef® is a hero in our menu packed with Country Comforts.

pasta & noodles

Vegetarian Marinara Fettuccine
Tossed in tomato marinara sauce with fresh garlic, mushrooms, spinach and feta
Add chicken breast or prawns

Farmer's Fettuccine
Noodles in a creamy garlic and parmesan sauce, with farmer sausage and mushrooms

Chicken & Broccoli Fettuccine Alfredo
Our creamy fettuccine, with fresh broccoli and charbroiled Fraser Valley chicken breast

Teriyaki Stir Fry
Fresh garden vegetables sautéed in teriyaki sauce piled over top Oriental noodles
Add chicken breast or prawns

At abc Country Restaurants in Canada, the menu is geared toward families. It prominently features Certified Angus Beef, which has helped the national chain build a strong following for beef entrées for breakfast, lunch, and dinner.

At the 1994 Canadian Food & Beverage Show, licensed distributors Macgregors Meat & Seafood and J.J. Derma work together to spread the word about Certified Angus Beef. Show participants include (left to right): Garry Lawson from Macgregors, Ron Chapchuck from J.J. Derma, export director Erickson, and Tim Chapchuck from J.J. Derma. Graham Newton (above right) from Macgregors joins the effort.

their beef. For some, U.S. beef was "absolutely forbidden," Riemann says. "They would never touch it." But Canadian packers producing the brand from Canadian cattle circumvented some of the image issues U.S. beef might have had at the time. In fact, it actually opened up some opportunity. A handful of Canadian retailers were primarily selling a different premium-beef product from the U.S. When the BSE crisis hit, the retailers picked up Certified Angus Beef because of its Canadian cattle connection.

Having non-U.S. packers might also improve international sales outside of Canada, says Geof Bednar, international director for Certified Angus Beef. When Hong Kong closed its borders to the U.S., Bednar says it continued to accept beef from Canada.

"There are always opportunities in some of our smaller markets outside of the U.S., where the Canadian product can be of added value," Bednar says.

While the packer initiative might have the largest impact on the brand, the licensing for Canada of the broadline foodservice distributor Gordon Food Service (GFS) in 2006 dramatically improved distribution options. The broadliner carries everything a restaurateur needs—from table settings to fresh foods to kitchen essentials. Some larger restaurants had switched all of their business to this type of distributor, a one-stop shop. Every day the brand did not have a broadline licensee, Certified Angus Beef was losing market share to a growing number of Canadian brands. Bednar saw a huge opportunity in the big picture, where GFS fit as a complement to the licensed meat specialty houses.

"Instead of ten salesmen, GFS had a hundred. And they handle ten thousand items," he says. "GFS helps us reach beyond major cities to surrounding communities and small towns."

On its Canadian website, GFS features its six top products that are "subject to the strictest quality controls." One of them is Certified Angus Beef, which the website says is "the world's premium brand of fresh beef. Exceptional taste." Notable licensees

In Canada, the President's Choice brand at Loblaws is synonymous with high quality. When that brand partners with Certified Angus Beef in July 2008, it gives 350 Loblaws stores a product that delivers authentic steakhouse flavor.

include La Bête bar-steakhouse in Quebec City, Restaurant L'Aurochs near Montreal, and Empire Grill in Ottawa, Ontario.

In terms of the black-hided cattle supply in Canada, vice president Larry Corah believes Certified Angus Beef is having the same impact as it did in the United States.

"In Canada, you're going to see a lot more black cattle," Corah says. "There are 25 to 30 percent today. Ten years from now, we'll look at Canada and see nearly twice as many black Angus cattle."

Mexico and Latin America

Until the early 1990s, nobody in the U.S. meat business wanted anything to do with Mexico. Sure, some purveyors sold product, but they were so skeptical of the market that they made their customers meet them at the border for a brief exchange of money and meat. This process, peculiar but legal, was not the most ideal way of doing busi-

In 2000, packers in Canada are licensed, helping to fill consumer demand in the market and some international destinations. Canadian-identified product helps distributors like Intercity Packers in Alberta and British Columbia market to the region's country-loyal consumers.

ness. Mexican customers were shorted on quality even though U.S. companies required them to pay in advance.

Without business relationships, Mexico became the dumping ground for off-size and second-tier quality products. Yet the people of Mexico and the steady stream of tourists seemed hungry for something better. Certified Angus Beef international director at the time, Tracey Erickson saw a growing middle class in the country, as well as a booming resort trade. She found a kindred spirit in 1992, when she contacted Houston entrepreneur Don Freedman.

Through his company, Freedman Food Service, he had made enough cold calls to hotels, steakhouses, and resorts in Mexico to agree there was a need for premium steaks in that marketplace, and an even greater need for a local supplier. After the North American Free Trade Act (NAFTA) passed in 1993, Freedman decided to create the needed infrastructure that would combine world-class service with the Certified Angus Beef product.

From day one in the development of his Mexican venture, today called ComNor, "America's premier beef brand was the centerpiece," Freedman says. ComNor still recognizes Certified Angus Beef as "*el mejor del mundo*," the best in the world, he adds.

In the early stages of business, Freedman consulted with his wife Martha's relatives in Mexico City. Meeting there on how to check credit, keep books, and understand government regulations and labor laws in the Mexican context, the relatives said not to worry about those things. Subsequently, the partnership of Comercial Norteamericana (ComNor) was born.

Freedman and his colleague, Ron Boatwright, now president of Freedman Meats, would ship Certified Angus Beef products, and the Mexican partners would handle the administrative issues. Skepticism was one initial obstacle to customers. Previously, no

When Neptune Food Service, a GFS Canada company, becomes a brand partner in 2006, it places strong emphasis on the brand. The British Columbia company redecorates its entryway (left) and displays the quality pyramid at food shows.

U.S. businesses had fulfilled any promise of a long-term commitment to the market. Still, foodservice customers such as steakhouses and premier hotels valued the concept.

When they saw a company committed to quality that could extend credit to creditworthy customers, and back its commitment by establishing local facilities, they gave Freedman a chance. He, Boatwright, and their Mexican partners met with customers in offices, smoke-filled back rooms, under umbrella tents on Cancun beaches, and in open air markets, from Mexico City and Monterrey to Cabo San Lucas. New relationships were being formed every day, and over time, the skepticism began to fade. It became clear that Comercial Norteamericana was there to stay.

While the concept of great service and product immediately took root, the partnership did not. "It didn't take long for us to set up shop and find out this was not a match made in heaven," Freedman says. "The Mexican partners wanted to exercise more and more control in areas in which they had no expertise. Less than a year into it, we had to part ways."

That left the two Americans stretched thin, but matters soon became even worse: In December 1994, the Mexican government devalued its currency. The peso quickly fell to less than half of its dollar value, more than doubling the price of a $10 steak from about thirty-four to more than seventy pesos.

The situation was so dire that the U.S. House of Representatives Committee on Small Business held a special meeting about the issue. In the hearing, Representative John J. LaFalce (New York, Democrat) said the devaluing of the peso could have "a pro-

In October 2009, Mexico-based distributor Comercial Norteamericana (ComNor) staff tours the United States. The company brings clients' chefs along for a meat science discussion at the Cargill Culinary Center in Wichita, Kansas. They also tour a Cargill packing facility, Angus ranch, and feedlot.

found impact on U.S. exporters and investors doing business with Mexico, especially the small business communities." Experts were predicting the peso to devalue an additional 10 to 20 percent, which it did as interest rates jumped higher.

For restaurants with bank debt, paying more than 100-percent interest furthered the threat, and many of Comercial Norteamericana's customers reacted by seeking out cheaper products. They hoped to return to the Certified Angus Beef brand, but the reality on the ground was a short-term fight for survival.

Freedman was further affected because the pesos that ComNor's clients paid would now pay less than half of the bill for the meat imported from the U.S. He remembers looking at the negative numbers and thinking, "What the heck did I get myself into?" In less than two years, he had suffered substantial losses and was wondering if things would ever be normal again. "If somebody came up to me and said, 'I will pay your accounts receivable and buy your inventory to take over your company,' I would have done it. But I knew in this economic environment there would be no easy solution."

The company hunkered down and demonstrated the level of its commitment to the market in Mexico. Perseverance paid off, and in fact, the currency crisis turned out to be a blessing in disguise. Freedman found a new set of partners who had run a successful meat business up until interest rates soared and the peso tanked. The Juan Manuel Najera family was in it for the long run, had a passion for the meat business, and had an equal passion for Certified Angus Beef.

"During 1995, the father, Manuel, and his sons Alejandro and Alberto took the bull by the horns," Freedman recalls. "As time went by, our confidence grew in them, and they led us into new markets. They were the right people."

Eliott Reyes Sr. (right), and his team at Provimi de Puerto Rico lead the brand's delivery to Caribbean islands beginning in 1986, when products are delivered by plane.

The Najeras helped the company grow from twelve employees to more than two hundred and fifty. By 2009, the Freedman team, through its subsidiary ComNor, had a huge presence in Mexico. It serviced more than fifteen hundred customers—"not just clients, but friends"—from seven centers in the country, selling some seven million pounds of Certified Angus Beef a year. If Freedman's company was a country, it would be the number four importer of Certified Angus Beef.

The modern Mexican market, built on Freedman's company's back, is a bright spot for the brand. O'Quinn says it's an emerging theatre with the potential for marketing fifty million pounds of Certified Angus Beef annually.

"Mexicans appreciate luxury brands and quality products, and we are recognized for both south of the border," she says. "The Mexican consumers are big carnivores; they love beef."

They also enjoy entertaining guests, O'Quinn says. "When they have a friend over, they treat you to the very best. They're just very big on showcasing quality."

Case in point: H-E-B Mexico, based in Nuevo Leon, licensed four stores in 2002. In just six years, it had thirty stores licensed and had become the top-volume small chain retailer in the world with fifty or fewer stores.

Numerous chefs and restaurateurs in Mexico are just as successful with the brand, and they are passionate about its quality for their customers. These locations include La Escondida and Hawaii Cinco Cero in Monterrey; Restaurante Loma Linda, Don Camilo, and La Vid Argentina in Mexico City; La Enramada in Cadereyta; and Hacienda San Agustin in Oaxaca.

Other carnivorous Latin countries with upside potential include Nicaragua, Guatemala, Panama, and Puerto Rico, which the brand considers as international business even though it's a U.S. territory.

H-E-B Mexico partners with supplier JBS Swift & Company for six-month billboard campaigns in 2007 and 2008 in Monterrey, the second largest Mexican city. In 2009, stores introduce consumers to the brand's deli meats and marinated chuck rolls.

The future of Puerto Rico looks extremely positive thanks, in part, to Reyes' company's legwork. A licensed distributor since 1986, Reyes has worked with the territory's culinary schools for ten years to supply the future chefs with Certified Angus Beef.

"We want them to handle and cook the product, so they can know it's available when they have their own restaurants," he says. Among his notable restaurant customers are long-time licensee Chez Daniel, known for its gourmet French cuisine.

Reyes also helps the Amigo grocery stores on the wine pairings that feature Gallo's Cabernet Sauvignon with various cuts of Certified Angus Beef. "The more people we reach, the more people who understand the difference between our product and others."

Certified Angus Beef depends on foodservice distributers, like Lombardi Brothers Meats in Denver, to maintain product quality as they age and portion-cut branded meat for restaurants. Licensed distributors are recognized meat experts in their markets.

7 Brand Evolution

By the mid-1990s, Certified Angus Beef was arguably a household name. It was served at such fine establishments as El Gaucho, John Q's, and Buckhorn Steak & Roadhouse, along with more than fifteen hundred others. Consumers could purchase it at large grocery chains like Giant Eagle and Giant Food Stores in Pennsylvania, Schnuck Markets in St. Louis, and Reasor's in Oklahoma. And perhaps most importantly, research indicated that Certified Angus Beef was increasing the price of Angus bulls by about $600 each. Moreover, the independent Nielsen Research Group found Angus was the most popular breed among the nation's cattle buyers and that packers viewed exotic breeds, specifically Simmental and Brahman, with "skepticism and distrust." Just ten years before that study was published, packers were buying as many exotics as they could because that's where the industry was heading.

Then, a producer-owned company began to evolve. It took some time and certainly some headaches, but once the ball was rolling, Certified Angus Beef was hard to stop.

The packing industry never looked back.

"What Certified Angus Beef does is bring that extra tier of top quality meat into the offering," says Charlie Mostek, senior vice president for Tyson.

Still, it's not like the brand hit the marketplace and boom, all the packers signed up. Certified Angus Beef and its licensees built the demand for this product, and packers have capitalized on its potential.

As consumer loyalty grew, distributors began requesting it. That's because consumers kept sending signals by purchasing more Certified Angus Beef. One of the good

Licensed retailers understand how Certified Angus Beef's assured quality and brand recognition are strong marketing points for their own brands. That's why they allocate valuable space in their most far-reaching tool, their circulars.

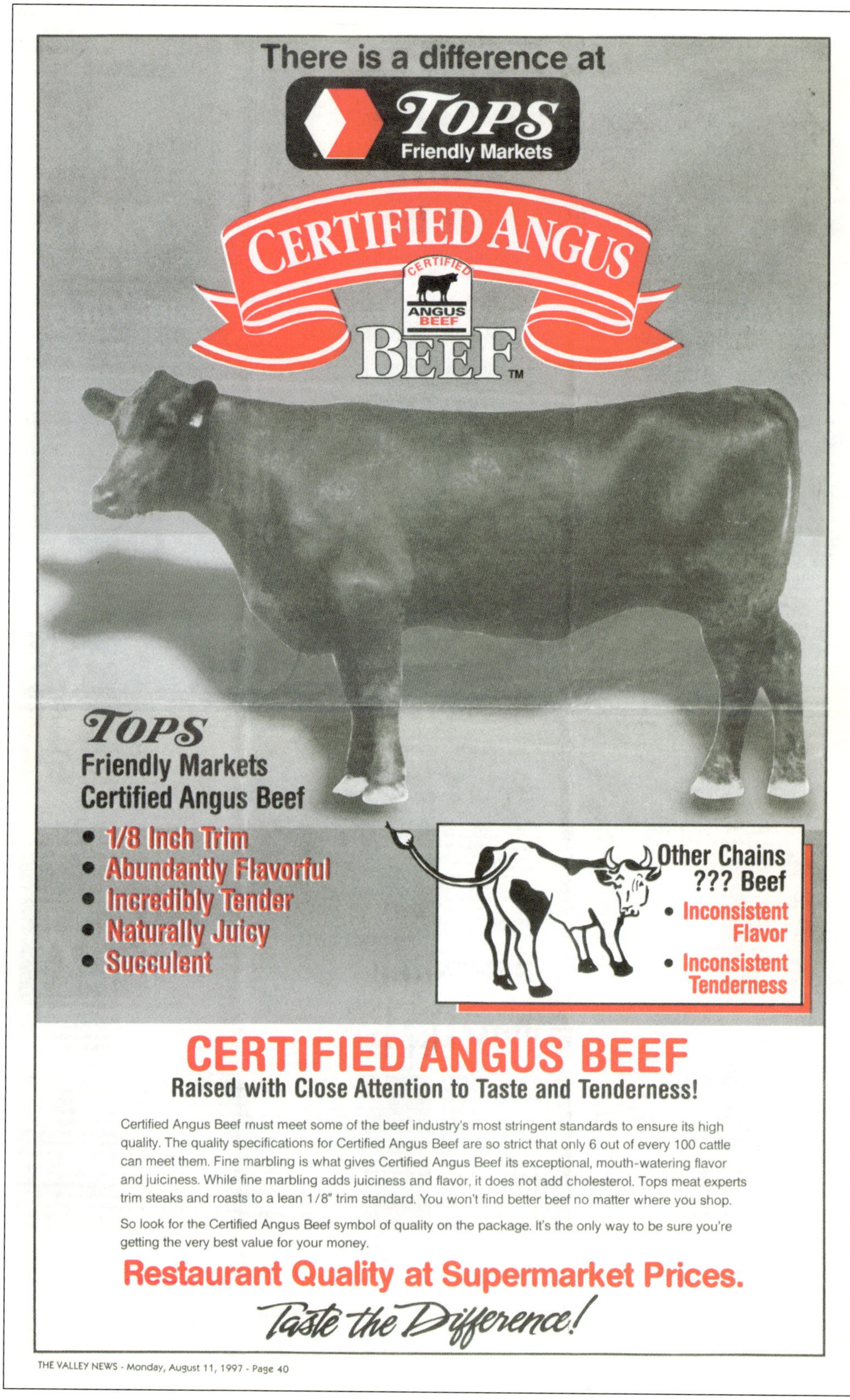

things about the pull-through demand was its positive influence on the way cattle buyers purchased animals.

In 1995, Tim Schiefelbein, then director of cattle procurement for Monfort, was consistently fielding requests for Certified Angus Beef product from his sales reps. But supply was limited.

"The best way I could get more qualifying cattle was to pay more for them," Schiefelbein says.

He started offering value-based grids, where cattle were paid for according to the carcass merit of each animal in a pen, instead of live-animal or dressed-weight pricing. Schiefelbein was the first to execute grid pricing at the company that is now part of JBS Swift. The tactic has since become very popular among packers. He believed it was the optimum way to procure for customer demands while rewarding the producer.

"It's all about value," Schiefelbein says. "The packer looks at what he paid for that animal. How much per pound on a carcass basis? How did that animal grade? What were the traits relative to that animal and how much value impact did each trait have on it?"

That was where the development of Angus genetics and feedlot management showed their true colors. While packers were licensing with Certified Angus Beef, many—even those who produced it—were still hesitant about fully committing because of the limited supply and low acceptance rates. Schiefelbein, who grew up in the Angus business, was proactive in finding the cattle his customers wanted, but not all packers were willing to take such bold steps to improve their supply line. To fill the void here, the brand's supply

The West Concord Super Market, in Massachusetts, has been licensed since 1983, the longest period for any retailer. This third-generation grocer enjoys loyal customers and builds brand equity with premium, hand-cut steaks.
Photo credit: Fred Minnick

Licensed since 1986, Roche Bros., in Wellesley Hills, Massachusetts, knows its customers well, making this beef-for-a-year promotion a hit.

development staff would encourage producers to collect carcass data and calculate expected progeny differences (EPDs) for carcass merit.

John Crouch, long-time Association official and retired executive vice president, said from 1974 to 1987, carcass EPDs were largely ignored. "Then producers realized that we are in the meat business and started paying attention to end-product merit," he said in a 1995 *Angus Journal* story.

Breeders began looking for bulls that had high EPD numbers in marbling, ribeye area, growth, and maternal traits that would continue to produce similar animals. In addition, bull owners were collecting carcass data from offspring in order to prove their animals' carcass merit. As available technology improved, so has the breeder's ability to improve the end product. Today, breeders are using DNA testing to show more information on genetic potential, and are using ultrasound to evaluate the beef quality of live animals.

Sam Johnson, owner of Summitcrest Farms and son of brand co-founder Fred Johnson, says his farm has been using the best available technology since 1985 to identify cattle that would qualify for Certified Angus Beef. "We have been on a mission," says Johnson, whose farm received the company's first Seedstock Commitment to Excellence Award in 1992. "If identified Angus cattle are qualifying for Certified Angus Beef at a 30-percent rate, we're looking for something better than that. Because of our use of carcass data, some DNA technology, and the statistical systems at the American Angus Association to identify the best genetics, we've got cattle that are producing upwards of 90- and 95-percent Certified Angus Beef—with the goal of putting those genetics into producing herds."

For Schiefelbein, a career cattleman and packer, it's nice to see Angus producers getting their due. He sees people paying $100,000 or more for high-marbling bulls, the U.S. herd turning black, and the incredible consumer demand for high-quality Angus beef produced with modern technology. It all started with the birth of a brand, he says. "It never would have occurred if Certified Angus Beef hadn't started."

Retail

While the packing industry benefited from improved Angus supply lines, the retail and foodservice segments benefited from more licensed packers. Simply put, packers are the gatekeepers of the meat industry. The more you can license to produce it, the more retailers and restaurants will carry your product.

In fact, a retailer that shares its hometown with Certified Angus Beef, repeatedly declined to carry the brand because the "supply line just wasn't there," says Bob Buehler, president of Buehler Food Markets. Many years later, in 1992, the Ohio grocery store chain would become a 100-percent Certified Angus Beef licensee. But the retailer couldn't commit "until a larger packer was brought on board."

Charlie Robinson, whose Oxford Trading Company was the first distributor, says Certified Angus Beef grew substantially in the retail markets with every packer it signed.

Retailers capitalize on the brand's Olympic suppliership at store level. Materials like this recipe card from Giant Food Supermarkets, based in Maryland, incorporate the supplier logo and capture American pride in the athletes and competition.

He specifically remembers the dramatic difference between the smaller packers and National Beef, when it came on board in 1983.

"The first five years of the program were tough, but as the packers came on, it became easier," Robinson says. "Now, we have competition, and competition sets the market. The prices started to adjust once they put on Excel and IBP."

The product was also more available through those larger packers, which helped Robinson sign more customers. One of those was the West Concord Super Market in Concord, Massachusetts, licensed since 1983, making it the longest running retailer in the Certified Angus Beef program. Located in a quaint village outside of Boston, this retail shop smells like mamma's kitchen, has the warmth of yesterday's diner, and is covered in Certified Angus Beef point-of-purchase material. Brothers Peter and Paul Mandrioli hand cut all the meat, know customers by name, and talk passionately about the brand.

"When we first heard about it, we wanted something that would make us stand out from the average store. We needed something different," Paul Mandrioli says. "Why am I sticking with Certified Angus Beef all these years? Because it's good. Time after time, the customers come in, and they just know what to expect. Certified Angus Beef never lets them down."

By 1986, larger grocery stores in Mandrioli's area were carrying Certified Angus Beef, but the competition seemed to improve their relationship with customers: "We're not really a self-service type of place. Most of our customers like to see it cut to order."

One of those larger retailers was Roche Bros., an eighteen-store chain based in Wellesley Hills, Massachusetts, licensed in 1986. Frank Vitale, the company's meat director, says at first, consumers loved the idea of a premium beef product, but didn't

1993

First packer offers ground chuck: IBP (Tyson)

First Canadian restaurants: Bardi's Steak House and Black Angus Steak House

Retail stores reach 1,000

Meat Processing article reports "Certified Angus Beef Sets New Standard;" in producer media, *Beef Today* publishes "In the Black" story

First distributors licensed in South America, Milenio Import & More; and Caribbean, Bahamas Food Services

Of 17 Angus bulls that sired more than 1,000 calves registered with Association, 13 have carcass EPDs (Expected Progeny Differences); seven have positive marbling and sired 11,746 registered calves

Thirty-three percent of nation's fed cattle have black hides

Top: Buehler Food Markets, in Wooster, Ohio, encourages shoppers to "Eat like a Champion" leading up to the Olympics. During the promotion, customers are randomly selected to receive their beef purchases for free.

Center: When Al Kober was named retail director in 2002, licensed grocers knew they had one of their own. Kober had spent fifty years in the retail meat business with Clemens Markets, in Kulpsville, Pennsylvania. He brings renewed energy and customer focus to retail, helping the brand and customers to new levels of success.

Bottom: Bi-Lo invites shoppers to grill their best for its first Certified Angus Beef Backyard Grill-off in 2007. Held during a street festival in downtown Greenville, South Carolina, the competition rewards the barbecue master with a new grill, trophy, and of course, bragging rights.

Training is a key strategy for sharing brand messages in stores. In 2006, Schnuck Markets holds its first Meat School for store-level meat staff, with sessions focused on brand quality and new promotional materials.

necessarily enjoy the extra price. "As the brand grew and people knew the quality of the product, a higher retail became less of an issue," Vitale says.

He remembers placing Certified Angus Beef in the company's service case and realizing how important the branded beef product could be to the meat department. More than two decades later, he says: "We've grown such a loyal following that if the product was removed, the consumer would let us know about it immediately. Your service case becomes a destination for premium products. At the holidays is probably one of the biggest rewards, because the shopper goes out of their way for a premium product. You get an opportunity to win that customer over for fifty-two weeks, not just the holidays."

Over the years, since Vitale has been a part of the program, he has noticed the dramatic shifts in the industry, but Certified Angus Beef has been a Roche Bros. constant because of its integrity, consistency, and the great marketing.

1994

Vic Stewart's first customer to receive National Beef Backer Award from National Cattlemen's Beef Association (NCBA)

Graycliff Hotel & Restaurant licensed; later longest continuous restaurant in Bahamas

Through USDA recommendation, beef muscling and neck hump evaluations transfer from live to carcass specifications

Value-added Products Division formed to increase chuck and round sales with specialty items like frankfurters and deli meats

CattleFax reports the brand contributes $354 to the price of an Angus bull

Certified Angus Beef Prime gives retailers, like Skogen's Festival Foods in Wisconsin, an additional point of difference in their full-service meat cases.

Like Vitale, Leonard Oddo remembers the up-and-down trends of the industry. As the service meat manager for grocery store chain Giant Eagle, he has worked with meat for most of his life.

In the late 1980s, Giant Eagle transitioned from selling USDA Prime at the service cooler to all Choice beef. Right away, Oddo noticed a drastic difference between the

Acme Fresh Markets, a northeastern Ohio grocery chain, is the first retailer to introduce Certified Angus Beef Natural on April 25, 2004. Owned by the Fred W. Albrecht Grocery Company, Acme carries a wide range of Certified Angus Beef fresh and value-added products.

meats just by handling them. They were tougher, didn't cut as well, and lacked substantial marbling. After selling the Choice product for a year, an independent Giant Eagle owner came up to Oddo and said, "We should be looking at another grade of beef," because he was losing loyal customers.

Giant Eagle started exploring new options. It was obvious the switch from Prime, while saving on costs, was a mistake. They had heard about Certified Angus Beef and invited Colvin to discuss the program.

"He talked about the cattleman's role, the production system, feeding practices, and all of the different attributes to Certified Angus Beef carcass identification," says Ed Steinmetz, vice president of meat and seafood for Giant Eagle.

The Giant Eagle staff liked what they heard and signed up for the program in 1991. They started in nine test stores and the product was a hit, Steinmetz says. But the initial reaction among managers was, "This won't work. Who will spend more money for this?"

As consumers did just that, Giant Eagle was adapting to the new program with its strict guidelines. According to Certified Angus Beef's "Partial Case Policy Parameters" sheet from 1990, retailers were eligible for the partial self-service case option only if they committed 100 percent of their service case to the brand. In the self-service aisles, several feet had to separate this brand from others, alternate protein sources had to be at least three feet on either side of Certified Angus Beef, and signage must be used to differentiate the brand. While these policies have since changed, Steinmetz appreciated them because they "made me feel confident that someone's making sure the specifications and the criteria are well managed and it doesn't get sloppy."

In a 1992 trade article, Colvin said some retailers were trying to take advantage of them. "They would widely advertise that they were selling Certified Angus Beef, but they

1994

HoneyBaked Ham Company licenses for catalog sales and stores nationwide

Thirty packing plants supply 172 million pounds of product, a 40 percent increase from prior year; nearly 10 percent exported

Federal ruling in lawsuit with Sysco Corporation supports brand's trademarks and quality assurance program

Office moves to new facility in Wooster, Ohio

Oklahoma retailer Reasor's ups the bar on beef quality in 2009. Licensed for 20 years, Reasor's launches Certified Angus Beef Prime Natural in stores, becoming the first retailer to offer this elite line.

would only allot us a very small portion of their meat case. We found we had to change our policy with regards to who could sell Certified Angus Beef in the retail case." The company wasn't trying to bite the hand that fed it—after all, at the time, retail made up 53 percent of its sales—but rules were rules.

The only rule Giant Eagle seemingly broke was very early on and it's doubtful that anybody caught it. An in-store advertisement said in bold capital letters: "Certified Black Angus Beef Is Now Available." Printed and distributed on April 19, 1991, Giant Eagle was well within its rights to publicize the brand, but the word "black" was a no-no. Looking at the ad now, both Steinmetz and Oddo chuckle. But in today's marketplace, the name is everything.

"Over the years, more and more Angus programs have been developed, and that creates some confusion," Steinmetz says. "The challenge on our part is to continually differentiate what we're selling from other programs that might be called Angus. It's not Certified Angus, Certified Black Angus Beef; it's not Black Angus Beef. It's Certified Angus Beef. You've got to continually police that so you can maintain the differentiation between all the other programs."

Seattle-based Quality Food Centers partners with the brand and Washington State Beef Commission in 2008 to sample new chuck steaks in stores and at Washington State University football games.

And sometimes licensees have had to police their own staff. Bob Boehringer, who has been a part of Certified Angus Beef since the mid-1980s, remembers when his Tom Thumb grocery chain first started carrying the product. Some managers tried to beat the system. "We had folks start to put out Select instead of Certified Angus Beef. We were like, 'What are you doing?' They would say, 'I can create some positive money here by trying to sell Select beef for Certified Angus Beef.'"

Boehringer says once customers started seeing the vast difference in quality between Select and Certified Angus Beef, "There was no fooling them. The customers started noticing the difference and knew what kind of quality they would be getting." Boehringer educated his staff about the program and parted company with the few who tried to switch out product.

At Tom Thumb, the 1990s were the golden years for Certified Angus Beef. "People wanted to buy it because it was a more marbled, tender, and flavorful piece of meat versus a regular Select grade of beef that we carried," he says. "To this day, Certified Angus Beef rib steak is still our number one item as far as driving sales. It's never been a huge burden to sell higher priced beef. It's always been an item that's done really well for us."

In 1999, Safeway acquired Tom Thumb, just nine years after Boehringer's company was acquired by Randalls Food Markets. In any business, acquisitions are just part of the game and typically the new parent company wants to share its line of products. That's what happened to Tom Thumb. It had to start carrying and featuring Safeway's in-house branded Select and lower-Choice program. While customers appreciate Certified Angus Beef quality, they have accepted the Safeway program, he says. Nonetheless, having two or more beef programs has become the industry norm. That's why when Al Kober became the program's retail division head, the first thing he did was eliminate some of the rules that wouldn't allow other brands to be near Certified Angus Beef.

"We lost some major accounts because we weren't willing to be flexible," says Kober. "We didn't want to change our integrity, but we had to modify some of the retail guidelines. The times had changed and we were not changing with them." Kober had worked fifty (yes, fifty!) years for Clemens Family Markets, a licensee since 1996.

In 2006, Certified Angus Beef staff develops new cutting methods for ribeyes and strip loins. They provide chefs and grocers with a solution for larger sub-primals, helping them serve thick steaks with an appealing portion size and plate presentation.

When he interviewed with the brand in late 2001, Kober wanted to know what could evolve in retail. He told the executive team: "The retail division needs to change its focus. It has been good up to this point. What can I change?" Corah, vice president and interim retail division leader, told Kober, "You can change anything but the specs of the beef."

As the interviews continued, Kober, a high-energy sort who runs to the copy machine and to his car for exercise, was the one asking all the questions. Riemann would say some years later to his friend, "It was like you were interviewing us and not the other way around." Kober was hired in February 2002 and the retail division, while doing

well, would get its biggest boost since signing Oxford Trading Company. Kober's hiring also marked the first time the company had a retail division director with retail chain experience.

"Hiring Al was the best thing Certified Angus Beef has ever done for retailers," Bob Buehler says. "I talked to people after they chose him and everybody was just ecstatic.

"When most people tell you they are in a hurry to get something done, they mean within a day or two. If you ask Al and you're in a hurry, he thinks that means one hour."

Buehler first met Kober at a Food Marketing Institute (FMI) conference, where Kober gave a dynamite address. "The guy is bigger than life," Buehler says. "Before he was with Certified Angus Beef, I heard him speak at FMI about meat advertising. I came back, tried some of his ideas, and they worked."

As Buehler says, Kober seemed to know everybody in the industry and was considered a leader. He would bring these contacts and his understanding of the industry to Certified Angus Beef. Under his leadership, the brand grew from thirty-three hundred licensed U.S. grocers in 2001 to more than fifty-six hundred in 2009.

The reason for this success is simple, Kober says. It's all about building relationships. "The mission statement we had in retail wasn't focused enough on the licensees' business. That changed," Kober says. "We are going to do everything possible to make their business successful. We needed to get more involved with the retailers, help them with their challenges."

One challenge for independent grocers and local chains like Buehler's is Walmart. With Walmart selling meat, every retailer has been trying to differentiate themselves from the big blue store down the street because, as Buehler says, "The days of loyal customers are disappearing fast. Walmart's a threat; it doesn't make life very pleasant when you're in independent grocery."

While not all licensed retailers have Walmarts directly across the street, many say the chain is a big concern and that Certified Angus Beef has become one of their major defenses against it. The brand has become a destination for customers.

In recent years, Walmart was selling Select and Choice beef pumped up with solution. But as Walmart and other retailers look for programs to compete with Certified Angus Beef, Buehler says, "We sometimes need to raise the bar," just like he did in the summer of 2008 with the brand's Natural line.

Summer brings peak demand for steaks. Ultra Foods, and its sister company Town & Country, capitalizes on this fact well. It offers quality beef at an incredible value to working-class Americans in Chicago's suburbs.

The National Cattlemen's Beef Association leads an initiative to match cuts with ideal cooking methods in its Beef Made Easy program. Retailers embrace the concept and use labels to showcase mealtime entrées, like this label for Certified Angus Beef at Giant Foods in Pennsylvania.

Since the 1980s, marketing support has gone beyond print materials to include menu design, waitstaff and chef trainings, and consultative sales. Certified Angus Beef partners with leading foodservice distributors to deliver superior-quality products, and they in turn work with restaurant customers to match the right product with each menu need.

"Natural has helped us with that group of people who come in and ask for organic, but they really just don't want it to have any added hormones. Natural will fill that need," he says.

Kober believes adding the Natural and Prime lines are two important steps helping retailers. After earlier test marketing with production from Tarpoff Packing in Granite

Certified Angus Beef specialists are a vital link between distributors and the brand. They energize, educate, and help guide their customers' beef selections. One specialist each year joins the elite group. Past honorees pictured with the 2009 recipient, Hubie Graham, are (left to right): Dave Trzesniewski, Ann Schmalzried, Mike Radis, (Graham), John Landry, and T.J. Leonard.

Specialists in the Field

Each year, one specialist at a licensed U.S. or international foodservice distributor is recognized for their dedication to the brand. The Specialist of the Year award recognizes that person for excellence in educating the sales team and local chefs, ensuring proper brand identity, and leading innovative promotions to drive brand sales.

"These specialists take their roles as brand ambassadors seriously," Polzer says. "They look beyond the day-to-day to ensure the commitment of their team and customers. They are focused on long-term success for both."

1999: Joe Boutte, Freedman Meats, Houston, Texas
2000: Jeff Savage, Malcolm Meats Company, Northwood, Ohio
2001: Mort Kent, formerly with Lombardi Brothers Meats, Denver, Colorado
2002: Danny Ransom, Intercity Packers, Vancouver, British Columbia
2003: John Landry, Sysco Kansas City, Kansas
2004: Mike Radis, Blue Ribbon Meats, Cleveland, Ohio
2005: Mark Idoni, formerly with Sysco Eastern Maryland (Lankford), Pocomoke, Maryland
2006: Ann Schmalzried, Sysco Detroit, Canton, Michigan
2007: Dave Trzesniewski, Sysco Eastern Wisconsin, Jackson, Wisconsin
2008: T.J. Leonard, Sysco Central California, Modesto, California
2009: Hubie Graham, U.S. Foodservice, Tampa, Florida

City, Illinois, Prime was formally launched in 2001 with supply assured by adding National Beef Packing. In 2004, the Natural product hit the market at B3R Country Meats in Childress, Texas. While internally there was a good deal of debate on the merits of creating a product outside the traditional line, retailers and restaurateurs quickly showed support for the programs.

For Quality Food Centers (QFC) in the Northwest, Certified Angus Beef Prime and Natural were accepted instantly because of the name. During a time when almost all food producers have "natural" lines, many people were skeptical of the label. "Among meat folks," however, "there's no other brand that people would trust more than Certified Angus Beef," says Oscar Blaser, meat director for QFC. "The brand is almost 100 percent of our beef sales, except for some commodity grinds."

QFC carries Certified Angus Beef Prime in the service counters, Certified Angus Beef traditional and Certified Angus Beef Natural. "When you're in a retail chain called Quality Food Centers, quality is the first word, so the meat in that case has to be of the highest quality you can buy. We don't come out and throw a USDA Choice or a Select steak on the advertisement, trick them into the store, offer them that; and they have an eating experience that's far less superior than what they would buy normally."

Blaser believes QFC consumers fully understand the value of the brand. "When I interview my meat guys, that's exactly what they tell me. Customers call us and say, 'Don't ever change your beef. I only buy the Natural; your competitor's meat is changing.' " In the downward spiral of the 2008–09 economy, Blaser and other meat managers felt the pressure of high food prices. "We all felt the pressures of cost increases. You have retailers out there who sacrificed quality to pick up some pennies and we're hearing that on the front line from random consumers and premium loyals. If you were to pull Certified Angus Beef out of our stores, our premium loyals would be raising hell."

Instead of cutting back, Blaser has been instrumental in finding new ways to add value to the product. In 2009, his breadwinner was still a twenty-year program of adding marinades and rubs to the packaged meats. Grocers have done that for years, but the difference with QFC's program is Blaser truly focuses on it, marinating what will be marketed that way just as soon as the fresh beef arrives.

"We're going to season the meat so that they don't have to worry about what to do with it, and we'll put the cooking instructions right on the package," he says. "It really takes the guesswork out of it. And the reason it's successful, too, is the consistency of Certified Angus Beef product."

More retailers are adding value through the new chuck cuts. For years, the chuck muscle had only been used for stewing, braising, or pot roasting. As the times changed and people had less time to cook a pot roast, the whole muscle or large cuts of it became less desirable. A few years ago, the beef industry began looking at ways to improve the value of the chuck. Recently, the Checkoff-funded Beef Innovations Group (BIG) came

Retailer's circulars are valuable marketing tools. Many, like Stop & Shop (left) and Shop Rite (right) in the Northeast, dedicate circular space to brand promotion and educating customers.

up with several new cuts, including the Sierra and Denver steaks from the chuck. BIG, whose board includes Certified Angus Beef technical services manager Mark Gwin, helped bring those innovations to end users. In March 2008, Meijer, the brand's highest volume retailer, launched four new chuck cuts at twelve stores in Cincinnati. QFC soon followed with a similar rollout.

"Back in the good old days, we'd put the chuck on the counter and it was gone," Kari Underly, a Chicago meat marketer, said in the May 2008 *Angus Journal*.

While those days are long gone, the new chuck cuts appear to offer new hope for retailers. Increasingly, they are taking the large chuck roast and stripping it down by each individual muscle, selling the more tender muscles as steaks, Gwin and Kober explain. Instead of selling the whole chuck for $3.99, they're selling the tender sections for $6.99 a pound. "The steak is as tender and flavorful as many traditional steak cuts, but less expensive. That's value for the consumer and retailer," Kober says.

Foodservice

The recently identified beef cuts offer similar value to chefs in foodservice, where brand evolution began at the licensee level. And there are few people affiliated with the brand more colorful and influential than Edd Hendee, owner of Houston-based Taste of Texas and a licensee since 1984. Hendee is also one of Texas' most-listened-to conservative radio talk-show hosts. In fact, he has been asked numerous times to run for Congress, but always says he can make a bigger difference through the airwaves.

In business since 1977, Taste of Texas saw its revenue grow 600 percent within five years of licensing. "Until we introduced Certified Angus Beef, we tried to be everything to everybody," Hendee says. But once he had dependable, high-quality beef, he focused on educating his employees about the program, promotions, and menu development. The restaurant's slogan, "Famous for Steaks," was more than just advertising copy. It was the truth.

Taste of Texas would become one of Certified Angus Beef's proudest stories. In addition to selling millions of pounds, Hendee has spoken at numerous events on behalf of the company, given tips to other licensees, and always has advice for company representatives. Perhaps more importantly has been the great experience Certified Angus Beef has given him.

"The first time I ever set foot in a packing plant or feedlot, I thought, this is really interesting and I began to see how it all ties together," Hendee says. "We began to have a relationship with Certified Angus Beef that we certainly have with no other supplier, and I don't think could ever have with another branded product. Our businesses have grown up at the same time."

As director of marketing concepts in 1985, Mary Ferguson recalls Hendee's "colorful" personality. He has always been a "brutally frank supporter," she says. In the early years, Certified Angus Beef provided licensees only with steak picks, a simple table tent, and two pamphlets; retailers got a couple of banners bearing the logo. The first time

Foodservice Industry Broadens

In the past few decades, the foodservice industry has been evolving to a relatively few large "broadline" companies, so called because they carry literally everything a restaurant might need from beef to pickles, napkins, and toothpicks. About 60 percent of the brand's foodservice distributors in the U.S. and Canada were broadliner owned in 2009.

"Specialty meat-cutting distributors remain the foundation of our brand," Polzer says. "They provide the portion-cut and aged steaks to their own customers as well as the broadline locations."

Since 1997, the new wave of broadliners has improved access to chains and accounts beyond major cities. They also serve a large base of hospitals, nursing homes, and schools, an emerging outlet for the brand.

Among the broadliners are three Foodservice of America locations in the Northwest and four Performance Foodservice locations, including longtime specialty meat cutter PFG Middendorf in Missouri. Sixteen U.S. Foodservice locations include U.S. Foodservice/Stock Yards in Phoenix, a meat house licensed in 1983 and later acquired by the broadliner. The category also includes numerous Sysco locations, taking in the licensed meat-cutting centers overseen by Malcolm.

"We look at it as our brand," Malcolm says. "They have a great product, we're proud to offer it, and we feel like it will only grow. There's a lot of respect between the partners."

Above: Distributors exhibit at food shows across the country, like this one with U.S. Foodservice Tampa, putting Certified Angus Beef on center stage. These shows help foodservice professionals find new products and ideas to grow their businesses.

Above left: Certified Angus Beef specialist Duke Dirksmeyer of Sysco Houston is a cheerleader for the brand in his market. He holds an "Inflate your Sales" contest to keep staff and customers focused on the brand. When a staff member achieves a personal sales goal, he or she throws a dart at the balloons to win cash.

Ferguson met the Texan, she knew what must be done. "Edd's comments provided the basis," she says. "I went back to the office and began developing marketing materials, policies, and programs that would aid licensees—both foodservice and retail—to promote products under consistent, proper trademark usage, and merchandising methods," Ferguson says. "Taste of Texas has gone on to be one of the most successful independent restaurants in the country, and Edd Hendee is one of the program's most avid supporters."

Loyalty in foodservice starts with distributors and thousands of sales representatives who coach their restaurants and help the Certified Angus Beef staff understand their issues. They must "know beef" to be consultants to restaurateurs. That is why Certified Angus Beef requests each distributor has a specialist for the brand. The specialist is the key contact for sales people at the distributor level and serves as a critical link to the brand's headquarters.

Since the specialist position's inception in the mid-1990s, it has proven to be an invaluable opportunity for distributors to increase their level of service to customers while essentially becoming brand extensions. In addition to driving Certified Angus Beef sales, the specialists train their licensees and promote brand awareness.

"I go around to our various accounts, talk about the attributes of the brand, try to get more Certified Angus Beef items on the menu, and try to get more accounts licensed," says Stowell, the specialist at Dole & Bailey.

Certified Angus Beef provides several training options for specialists because their role is so critical. For example, Hussman frequently comes across restaurants that want to buy the product, but not become licensed and carry the logo. In one sense, he wants to help Certified Angus Beef optimize its licensing potential; in another, he must meet the client's needs or they'll go somewhere else.

Leonard Berkowitz often consulted with the likes of Newport Meats to help sort out these very issues. After him, it was Arnie Hein, from long-time licensee

Opposite: Distributors have a way of making Certified Angus Beef their own, motivating their sales forces to reach new customers and sell new items. Southern Foods, licensed since 1986, posted this "3 Million Pound Drive" poster, hoping to achieve its sales goals and perhaps annual conference recognition.

Southern Foods, who consulted after retiring in 1997. "I worked with established distributors that wanted help in one way or another," says Hein, who served on the brand's board in 1999 and 2000. "I mostly gave presentations to sales people about Certified Angus Beef."

On the not-for-profit company's dime, Hein helped distributors with plant operations and business practices. "In one instance, I was asked to come in and referee between three brothers in a family dispute over who was going to succeed as president of that company—the older brothers were past retirement age and the younger one wanted to take over. They didn't quite know how to get that done."

During Hein's consultancy, a rising star at Freedman Foodservice in Houston was setting the world on fire. Joe Boutte grew up in the business, watching his father cut meat at his butcher shop. Boutte then became a professional himself. In 1998, he was named the first Certified Angus Beef Specialist of the Year.

Boutte says winning the award was a defining moment for his career, but the product did most of the work. "I did taste comparisons," he says. And the brand won people over every time.

Smithco Meats in Seattle operates in a similar fashion. When Patrick Marshall of Smithco and his wife attended a Cirque de Soleil dinner theater, they were served New York strip steaks. "It was absolutely foul. The beef ate like a shoe," Marshall says. "It had a bad flavor, and it was as tough as it could ever get—and it was not overcooked."

Marshall, a Certified Angus Beef product specialist, saw it as a selling opportunity. He called the operation and learned they had about a 15-percent return rate on those steaks from free-range, grass-fed cattle. Marshall sold them on Certified Angus Beef flat iron filets and since then, the company has not had a return on the item.

"The joy of carrying this product is you don't get the calls, as we do when we sell any commodity beef," Marshall says. "Somebody'll say, 'My guests were kind of complaining about the toughness.' Well, that's because you insisted on the lowest-price commodity beef we can possibly put into your kitchen. Let's talk about quality and consistency. Here's something else that we carry."

That "something else" is Certified Angus Beef. As Hussman alluded to, many restaurants in the Smithco market don't want to license with the brand; they just want to carry the product. But since 1997, Smithco has licensed some very prominent accounts, such as El Gaucho, with its focus on independent restaurants.

A big challenge for Smithco comes from the smaller, local programs.

"There are so many Washington brands out there, and a lot of them don't have a big supply line," says Bob LaGassey, president of Smithco. "It seems some of them are killing as few as fifty head a week. Well, you can't put a whole lot of customer base on board with that low of number. The nice thing about Certified Angus Beef is that it does have a good packer base. You have a better opportunity to get product, fill your orders, and keep people on the program rather than saying, 'Gee, I promised you a bill of goods, and now I've got a thimble full of meat here to try to fill your order.'"

SHARE the VISION

You can help

3

SINCE 1978
CERTIFIED ANGUS BEEF®
BRAND

Million Pound Drive

* Southern Foods

One of a distributor's best marketing tools is its delivery trucks. Licensed distributors want their restaurant customers and consumers to know they proudly deliver the best beef available. A moving billboard is an invaluable and effective promotional tool.

LaGassey says Certified Angus Beef also helps him over the competition with an arsenal of product information and marketing material. "You don't have to make up the story; they've already pretty much got it for you," he says. The brand also started reaching out to the restaurants, offering videos and literature filled with helpful selling tips for use in waitstaff training.

"If waitstaff explains the product to customers and essentially sells more, it's more profitable for that restaurant," says Deanna Walenciak, former assistant director of the foodservice division and current director of trade-level marketing. "Beef items are

In 1997, Certified Angus Beef introduces a national contest for waitstaff to help them talk about the product's quality with customers. Company president Jim Riemann (left) presents the Horse and Hound Restaurant with the grand prize in 1999. Accepting are Beth Rose and Lola Getzlaff (left to right).

more profitable than poultry or pasta, so training helps the restaurant drive sales, and we have another avenue for consumer education."

Perhaps the number one reason restaurants work with Certified Angus Beef distributors, however, is the level of service they provide. Brand specialists are knowledgeable professionals who help them with improving customer satisfaction, word-of-mouth advertising, and ways to become more profitable. Considering the restaurant business can be volatile, most restaurateurs will take expert advice anywhere they can find it.

Lee Fritz, Certified Angus Beef specialist and vice president of sales for Neesvig's, says his Wisconsin-based meat distribution company is always trying to help build his customers' business. His assistance was especially appreciated in 2008–09, when the restaurant industry faced its toughest time since the late 1970s. When customers came to Fritz about cutting costs in their beef program, he would help them carry less-expensive specials like flat iron steaks or tasty medallions made with unique sauces. But he never once fell into the mantra that it was time to switch to Select beef to cut costs.

"In trying times, the guy that stays with quality tends to weather the storm a lot better than the guy that does nothing but cut costs," Fritz often advised. "We're trying to be that valuable supplier that helps them keep a diverse menu that brings in profit. They've got to keep afloat so that we can all stay in business."

Fritz also extended the full line of services Certified Angus Beef's marketing department offers, which includes advertising resources, training, and menu design. In 1998, the brand began designing menus for customers, with the first menu developed for Charlie's Chop House in Southgate, Michigan. "That was another opportunity to ensure proper brand identification on menus and provide service to licensees that no one else can provide better," Walenciak says.

Saltgrass Steak House, based in Texas, strives to "honor the old legends, while makin' a little history of our own." In 1996, it begins serving Certified Angus Beef at seven locations and promotes the brand at the Houston Livestock Show and Rodeo. Today, it has forty-three locations in five states.

Opposite: Uno Chicago Grill is known nationwide for its legendary Chicago-style pizza, and in April 2007, the chain adds Certified Angus Beef steaks to the menu. Its product rollout includes menu promotions as well as print, radio, and television advertising.

While licensees appreciate those services, many point to the Prime and Natural extensions as the best decisions ever made by Certified Angus Beef. But brand executives were quite cautious about moving forward with new lines. The question with the Prime line was, "Are we saying our traditional product isn't good enough?" However, Tarpoff Packing Company in Illinois was already segregating and selling Certified Angus Beef Prime at a higher premium. The licensed packer had done the market research.

DO YOU KNOW YOUR
STEAK?
HERE AT UNO CHICAGO GRILL® WE TAKE PRIDE IN OUR 60 PLUS YEARS OF EXPERIENCE AND DEDICATION TO THE CRAFTSMANSHIP AND QUALITY OF ALL THAT WE DO. THIS IS WHAT SETS US APART FROM THE REST. WE'RE CONVINCED OUR STEAKS WILL MOVE YOU AND WE KNOW YOU'LL COME BACK FOR MORE.
STEAK BASICS 101
We feature selected cuts of the Certified Angus Beef® brand
Certified Angus Beef® steaks are recognized for consistent flavor and tenderness
Only 8% of all beef qualifies for the Certified Angus Beef® brand
The marbling of beef is the most important Certified Angus Beef® qualification
THE BOTTOM LINE
At Uno Chicago Grill we offer premium quality beef.
We feature selected cuts of the Certified Angus Beef® brand–the best tasting beef available.
But why listen to us, TASTE IT FOR YOURSELF!
BEEFOLOGY Q&A
Q: WHAT GIVES BEEF ITS FLAVOR?
A: The flecks of fat within the beef, marbling.
FACT: Marbling gives beef its juiciness, flavor and tenderness.
Q: IS BEEF HEALTHY?
A: Yes! Beef offers significant nutrients as part of a healthy diet.
FACT: Beef is the number one protein in America. Beef is a significant source of zinc, iron, and protein, as well as B-vitamins. Calorie-for-calorie, beef is one of the most nutrient-rich foods to fuel an active and healthy lifestyle.
SINCE 1978
CERTIFIED ANGUS BEEF® BRAND
NOW FEATURING SELECTED CUTS
HAND HARVESTED SEA SALT
WE SPRINKLE ON OUR FLAVOR ENHANCING SEA SALT CREATED FROM THE WARM WATERS OF THE MEDITERRANEAN SEA. MADE FROM THE ENERGY OF THE SUN AND WIND, THIS FINEST QUALITY SEA SALT IS STILL PRODUCED IN THE ANCIENT PHOENICIAN TRADITION, RESULTING IN A SUBTLE PURE TASTE...
NOW YOU KNOW YOUR STEAK AND YOU LIKE IT
STEAKS
GRILLED SHRIMP & SIRLOIN*
Tasty half-pound of Certified Angus Beef® Top Sirloin and five skewered shrimp.
THE CHOP HOUSE CLASSIC™*
10 oz. of Certified Angus Beef® Top Sirloin served with a House or Caesar side salad.
TOP SIRLOIN*
Certified Angus Beef® tender and rich in flavor. Available in 8 oz. or 10 oz.
UNO
7 oz. FILET*
Deliciously juicy, tender and flavorfully cooked to your personal perfection.
All steaks served with your choice of side and a breadstick.
*Items marked with an asterisk may be cooked to order. Consuming raw or under-cooked meat, poultry, seafood, shellfish or eggs may increase your risk of foodborne illness, especially if you have certain medical conditions.
FOR MORE SIZZLING STEAK, GRAZE OUR OTHER SELECTIONS IN THE MENU...

In 2007, Food Services of America-Boise paved the way for Certified Angus Beef to become the official branded beef for the Roady's Humanitarian Bowl. Hans Schatz (left), from Foodservices of America, worked with Certified Angus Beef licensee John May, owner of The Gamekeeper in Boise, on the Bowling for Beef event (pictured) and the bowl game's official dinner.

President John Stika says once people looked at the science of Prime it was accepted. "The value of Prime is defined by science, but the same cannot be said for Natural. The staff again voiced some apprehension—what does it say about our other product? Should we offer it under another brand name?"

Those questions soon subsided after enough foodservice and retail distributors said, "We need Natural." Internally, the Natural line is called the "never, never, never program" because the animals never receive antibiotics, hormones, or animal byproduct feeds, and are always traceable to their origin.

"We think Natural is going to continue to grow," says Andy Malcolm, chairman of Sysco Specialty Meat Companies. "I think people in the industry are reaching for any opportunity to do things a little differently, and know the consumer is looking for that kind of product much more today than they were ten years ago."

Restaurant chains are another area of growth for the brand. Fourteen years after Old San Francisco Steak House became the first licensed U.S. chain, Certified Angus Beef licensed Chili's, a five-hundred-unit chain, to carry a country-fried steak. In place for a number of years, that was the first agreement with a major restaurant chain and one of the first one-item licenses. Since then, Hard Rock Café, Cheesecake Factory, Uno Chicago Grill, and J. Gilbert's have followed suit. Some regional chains like Saltgrass Steak House started serving the brand at just a few locations, and after building the business around it, Saltgrass expanded to forty-three restaurants in five states. Tripps, based in North Carolina, has also grown with the brand to fifteen locations and added its Rock-ola concept.

It may seem ironic that Certified Angus Beef was built by independent restaurants like the Taste of Texas, but its future embraces the chains. For Mark Polzer, vice

Home of the NHL's Phoenix Coyotes, Jobbing.com Arena serves Certified Angus Beef and sports the beef icon's logo. U.S. Foodservice Phoenix has built a strong partnership with arena officials and helped brand its Chuckwagon Grille concessions.

Via Christi Medical Center licenses in 1999. This holiday promotion let patients and their families know their cafeteria offers some great comfort food in Certified Angus Beef.

Via Christi Regional Medical Center

Licensed in November 1999, the Via Christi Regional Medical Center, in Wichita, Kansas, became the first 100-percent licensed medical facility.

It all started when Patty Dollarhide, former director of Food and Nutrition services, began seeing pizza boxes stacked outside of patients' rooms. She realized patients and their families were going elsewhere, even though the hospital offered well-balanced, nutritious meals.

She told the *Wichita Business Journal* in 2000 that the hospital's food was losing its appeal to the people who needed it most—those recovering from illness and surgery.

"Every day we're battling that old stigma of 'hospital food is bad,'" said Dollarhide. "So even if you prepare the healthiest diet for your patients, if it's not tasty and appealing, patients just won't eat it."

Sysco-Kansas City, a licensed Certified Angus Beef distributor, set the hospital up on an account using several underutilized cuts and two traditional steaks in its cafeterias and for patient meals. Doctors raved about the high-quality beef program. To introduce the program, they held a "steak day."

Wood Ranch BBQ & Grill, a licensed chain in southern California, shares the spotlight with the brand in 2009 during the premier of Disney's Hannah Montana *movie. More than a thousand guests attend a red-carpet event, where Certified Angus Beef mini-burgers are served.*

president of business development and former foodservice director, it's not a matter of irony. It's just a sign of the times.

A decade ago, on his way to work, Polzer would see six restaurants. "Now there's twenty-three, and all but two are national chains. We've seen, up until 2008, tremendous growth in the restaurant industry, a high percentage of which has come through national chains," he says. "If we want to continue to drive sales in the foodservice division, we're going to have to play with the chains, too."

Among chefs, Certified Angus Beef has become the most dependable, consistent branded beef available. They rely on its flavor, juiciness, and tenderness to please their guests.

8 Chefs and Owners Love the Brand

DeBragga and Spitler, the longtime licensed New York City distributor, cooked steaks for John Doherty when he was executive chef at the Waldorf=Astoria. "It was love at first bite," Doherty says. Soon after, he switched from USDA Prime to Certified Angus Beef. But there was some push back from the hotel managers.

The management challenged Doherty on the switch, because the hotel's Bull and Bear steakhouse must have the best. "Certified Angus Beef tastes better than USDA Prime," he'd tell them. They would typically reply, "Ah, no way, it can't be true."

To prove his point, Doherty set up blind taste tests. His food and beverage director and manager would eat tasting portions of both steaks. When they were done, they pulled out a piece a paper from under the plate with their favorite steak. Every time, the piece of paper said Certified Angus Beef.

As managers came and went, Doherty held the blind taste test several times over the course of two decades because doubters persisted. In 2002, the Waldorf=Astoria began serving Certified Angus Beef Prime. "When that came out, everybody got off my back. With Certified Angus Beef Prime, the business started to grow in leaps and bounds, like 20 percent in a year. People were dumbfounded at the quality of this steak."

That's high praise coming from one of the most influential chefs in the United States. Doherty has cooked for six U.S. presidents, countless celebrities, and royalty from all over the world. His book, *The Waldorf=Astoria Cookbook*, received rave reviews from national critics, and Bull and Bear is featured on *FOX Business News* every day,

Mark Miler, a James Beard award-winning chef, is among thirty-six chefs who share recipes in The Art of Cooking with Certified Angus Beef *cookbook in 1989.*

Right: In the 1980s, the brand gains popularity with chefs of renowned restaurants. George Wells Meat Company licenses the Old Original Bookbinder's in Philadelphia, a gathering spot for presidents, international dignitaries, and celebrities.

live from the bar to start the dinner hour. In his nearly immaculate kitchen, Doherty oversaw seven chefs de cuisine, ten sous chefs, and 110 cooks.

"He's an amazing chef," says Marc Sarrazin, owner of DeBragga and Spitler, who has worked with Doherty for thirty years.

Now a principal in Wolfpack Hospitality, Doherty has been a kind of cheerleader for the Certified Angus Beef brand over the years. He's become an official brand ambassador, sharing his enthusiasm with such media as *Food & Beverage Journal*, *About.com*, and *Catering Magazine*. In 2008, he starred in a video about the brand and was featured in an *Angus Journal* story. To capture footage for the video, he visited Sitz Angus Ranch in Dillon, Montana, and it was viewed at a media reception in the Bull and Bear to celebrate the brand's thirtieth anniversary.

The appreciation goes both ways and comes to mind whenever the brand considers a new product initiative. "We ask ourselves what John Doherty would think about this," says John Stika, Certified Angus Beef president.

Doherty's passion for the product is simple: the natural flavor.

"Even without seasoning, it is the most flavorful beef that I have come across. And, it's extremely consistent. It's really that simple. It tastes great all the time," Doherty says. "What makes Certified Angus Beef product consistent is the amount of energy and attention that's put into every step of production, from the breeding to raising the steers to the way it's handled. By the time it gets to me, the only thing I can do to screw it up is overcook it."

Many restaurant owners and chefs agree with Doherty. Rick Cassara, owner of John Q's Steakhouse in Cleveland, has used it since 1987. "We have never had bad product," Cassara says. "Certified Angus Beef is the most consistent product out there."

Chef John Doherty (right), the executive chef for twenty-three years at the Waldorf=Astoria in New York, takes a few days from his kitchen to learn about the brand from gate to plate. He visits with Donna Sitz-Arthun, matriarch of Sitz Angus Ranch in Montana.

John Q's waitstaff presents a tray of steaks to diners, explaining the flavor, quality, and importance of Certified Angus Beef. Cassara says consumers understand the brand and appreciate the quality associated with the name.

"You see an awful lot of imitators out there, but we've not found anything better than Certified Angus Beef," he says.

Like John Q's, El Gaucho steakhouse has the old-fashioned feel with waiters in tuxedos, and it serves Certified Angus Beef. Paul Mackay, owner of the Seattle-based restaurant, was amazed when he tasted the product. His first sampling came when he was researching meat purveyors.

"When I tasted it, I said 'my God!' I got on the phone with Bob LaGassey of Smithco and said I wanted an exclusive. It was so good!"

Mackay opened in 1996 carrying only Certified Angus Beef. He encouraged employees to understand the product by talking to one another about it and eating no other

Photo credit: Fred Minnick

So Long Chef

Paul Mackay encourages his employees to make decisions in the kitchen. Once, he gave a chef some leeway in cutting production costs in reference to the amount of green beans, risottos, and cheeses served. Mackay wanted him to cut back on waste.

Well, the chef took that too far. For about two weeks, he had stopped ordering Certified Angus Beef from Smithco and was purchasing Select and lower Choice from a nearby grocery store. Mackay was tipped by a cook who said, "There is some commodity beef in our house that we're serving."

"The chef was using that to cover his costs. I'm not unreasonable. I'm a fair guy." But Mackay has to maintain the consistent quality in his steakhouses, and if he cannot trust the head of the kitchen, "that's really scary," he says. "You don't cut back on the most important item in our restaurant," especially without the owner's permission.

In the 1990s, Certified Angus Beef sponsors a culinary competition with the American Culinary Federation. Hartmut Handke (left), a Certified Master Chef, receives his winnings from staff member Jenny Stickley in 1993.

Competitors in 1994 are pictured with staff member Lori Franklin. They are (left to right): Victor Gielisse, a Certified Master Chef; Handke; Stafford DeCambra; and Tom Windorfer.

meat when at work. Mackay says it was important to get his employees to "believe" in the program, because it wasn't Prime and nobody else had it.

"We got better value for our guests with this beef than we do with the commodity Prime," he says. Subsequently, Mackay stepped up to include the branded Prime steaks on his menus, and he has El Gaucho locations in nearby Bellevue and Tacoma, as well as Portland, Oregon.

1995

First seminar for brand "specialists" with licensed foodservice distributors

Blake's Lotaburger licenses 77 restaurants

"Taste the Difference" created as multi-media advertising kit for restaurants and retailers

Meijer licenses 14 locations, and becomes top volume retailer by 2006 with more than 170 stores

This braised chuck roast recipe was created by Chef Fritz Gitschner, of the Houston Country Club. Certified Angus Beef and Freedman Food Service sponsor his participation in the Bocuse d'Or world competition in 2005. A certified master chef, Gitschner offers the brand on the country club's menu.

He aims for the average diner to spend $100. That's a lofty goal, but Mackay jokes that he has to "provide a place for people to part with their money." However, those dollars would likely stay in their pockets if they didn't like the steak at his premier steakhouse.

"Certified Angus Beef gave us an anchor that I could rely on so we could take care of other things, such as the liquors, staff, and atmosphere," he says.

Certified Angus Beef corporate chef Scott Popovic has worked at many great restaurants, including N9NE in Las Vegas' Palms Casino, Aureole in Las Vegas' Mandalay Bay, Magnolia's in Charleston, and Fire Food and Drink in Cleveland.

Culinary Emissaries for the Brand

In fall 2003, the marketing team saw a need to make more direct connections to consumers in targeted markets. A corporate chef could become the lead spokesperson for the brand at consumer events and media appearances.

"Consumers associate with chefs in their markets and on food television," says Tracey Erickson, vice president of marketing. "Our corporate chef helps them understand why this beef tastes so good and provides them with easy meal ideas."

In January 2004, corporate chef Dianna Stoffer shared recipes at home-and-garden shows in Pittsburgh and Detroit. She prepared steak and eggs for local disc jockeys, while inviting radio listeners to the food shows.

The brand's chef also reaches out to other chefs and restaurateurs, talking in culinary terms about the product's specifications and why it is good for business. Outreach includes creating recipes for new beef value cuts, like the teres major, flat iron, and ranch steak, which are profitable for the restaurant and tasty for the consumer.

"I can tell you how to cook the meat; how to make it flavorful; how to make it exactly what you're thinking of; and I can tell you why I'm using the Certified Angus Beef brand," says Scott Popovic, who became the corporate chef in August 2006.

Since he joined the team, audiences have expanded to include Angus cattle producers. Demonstrations help producers understand the brand's focus on consumer marketing and product merchandising, efforts that create more market share for their brand and cattle.

When he's not in the kitchen, Popovic is hobnobbing with chefs. Very well-connected, he has the numbers programmed into his cell phone for a dozen celebrity chefs, an Iron Chef, and owners of multimillion-dollar restaurant concepts. These connections led him and the brand to the James Beard House in New York in October 2008 and the Share Our Strength fundraising barbecue at the National Restaurant Association Show in May 2009.

"Chef Scott delivers the credibility we had hoped to achieve with this program," Erickson says. "Additional staff including chefs Michael Ollier and Kyle Miller now work with him to maximize our outreach to both chefs and consumers."

1995

Licensing Monfort (JBS Swift & Co.) increases access to 70 percent of nation's fed cattle

First distributor in Japan licensed for retail: Sumikin Bussan Kaisha

USDA presents brand with Industry Recognition Award, for pioneering quality marketing systems for beef

More than one billion pounds sold to date

New export markets include China and military bases in Europe and the Middle East

Of top 50 Angus bulls registered, 39 have carcass data and sired 45,470 calves

Café Escadrille executive chef Brian Theurer shares that sentiment. He has been using Certified Angus Beef in his Boston restaurant for more than twenty years and regularly has five or six items on the menu. "It's very consistent," he says. "I never have to worry about it. Why change?"

Trends

The future of Certified Angus Beef somewhat depends on the various trends occurring in restaurant kitchens. The restaurant industry sees trends come and go. And nobody knows that better than Doherty. Through thirty years, he says, beef has remained ahead of chicken and pork.

"We've seen three recessions and various fluctuations in the economy. And beef still is, by far, the number one consumed protein. Beef has continued to grow in demand and popularity," he says.

One beef trend that has come on very strongly is the promotion of secondary cuts versus the filet or strip loin. Doherty says operators are braising more with cuts from the chuck and short ribs, to reduce costs. Since 2006, Certified Angus Beef executive chef Scott Popovic has been capitalizing on that trend.

"Chuck roll cuts are coming to the forefront," he says. "We're taking the boneless chuck rib and playing with different cooking techniques to see how it tastes."

Opposite: Rick Cassara, owner of John Q's Steakhouse in Cleveland, operates one of Ohio's finest. Open since 1979, its menu features several Certified Angus Beef steaks, including porterhouse, strip steak, bone-in filet mignon, and bone-in ribeye.

Left: Founder of the Mackay Restaurant Group, Paul Mackay is a restaurant mastermind. His El Gaucho locations in Seattle, Bellevue and Tacoma, Washington, and in Portland, Oregon, serve dry-aged Certified Angus Beef Prime cuts.
Photo credit: Fred Minnick

As restaurateurs look for value, the chuck cuts are extremely attractive because they're cost effective.

The Waldorf=Astoria is sticking with the middle meats but it carries two tiers. Diners can choose between Certified Angus Beef or its Prime brand extension for two different price ranges. Typically, the result has been a fifty-fifty split, the hotel reports.

In terms of preparation, many chefs are putting rubs and marinades on steaks with unique ingredients such as pomegranate, tamarind, garlic, espresso, and smoked paprika. Popovic believes there will be an upswing of Mediterranean, African, Jamaican, and Scandinavian spices. But not at John Q's.

Cassara, like many steak enthusiasts, has always believed in letting the meat do the talking. Only John Q's top sirloin, which is grilled, topped with melted bleu cheese, and served with a red wine demi-glace, is created with more than salt and pepper. Although he's noticed the evolution of steakhouses, and his cooks will prepare specialty steaks upon request, Cassara thinks a good steak is a good steak. And it doesn't need anything more than a little salt and pepper. "I just think that's the way you eat a steak."

Natural, organic, and grass-fed have become trends in the past ten years. With the introduction of its Natural product extension, Certified Angus Beef actually captures many restaurant consumers looking for organic because they just want to avoid any added hormones or antibiotics as the cattle are raised. As for grass-fed beef, many chefs are jumping on the trend, but it requires longer aging and typically a sauce because

1996

Country House Restaurant licenses; later longest continuous restaurant in Asia

First distributor in the Middle East: Aramtec

First research identifies consumer awareness, recognition, and perceptions of brand

Sizzling steak photo and consumer brochure developed for Japan; materials also offered in Spanish and Mandarin Chinese

USDA sets GLA standard for all brands with Angus influence as cattle source

Top: There is no greater honor for a chef than cooking in the James Beard House. For Certified Angus Beef's thirtieth anniversary, corporate chef Scott Popovic recruits leading chefs to prepare a meal for VIPs. From left: chef Randy Sebastian and Irina Dan, Rio All-Suites Hotel & Casino, Las Vegas; sommelier Manuel Nieves, Food and Beverage Solutions, Cleveland Heights, Ohio; chef Jerry Weihbrecht, Zoe's, Virginia Beach, Virginia; chef Michelle Brown, Jag's Steak & Seafood, Cincinnati; chef Dino Jagtiani, Rare and Temptations, St. Maarten; Popovic; and chef Cedric Tovar, Waldorf=Astoria, New York.

Photo credit: Erin Gleeson

Bottom: In May 2009, Certified Angus Beef participates in the 11th Annual Championship Cookout at the National Restaurant Association Show in Chicago. More than 1,100 industry influencers attend the fundraiser for Share Our Strength, the childhood hunger organization. Chef John Doherty is interviewed at the event for The Food Network.

there is little marbling. "The grass-fed is not going to eat as well as a grain-fed beef," Theurer says. "People come out to eat and want a nice, tender steak. They won't find it in a grass-fed."

Mark Polzer, vice president of business development and former sales manager for Becker Foods in Milwaukee, says chefs will occasionally get caught up in the hype of grass-fed, but when it comes down to it, the product better perform. "You're not going to sell a chef on smoke-and-mirrors. It better taste good. That's why they put grass-fed on the menu for a month. Then they take it off."

Then, there's the Kobe beef trend among fine-dining restaurants. Cleveland-based chef Doug Katz says Kobe is more of a short-lived fad. "Because, I don't think they're doing the same things that they are in the Asian countries with the Kobe," he says. "It doesn't compete with Certified Angus Beef. We use a rib steak and it's just so tender. Even with the filets, where fat wouldn't be an issue, I tend to go with Certified Angus Beef."

Doherty says, "Kobe beef is just a product on its own. It's extremely rich. If you want to have a steak, you're not going to have Kobe beef. You're going to have Certified Angus Beef. If you want a really rich, decadent slice of beef, you can go for the Kobe. It's just like eating foie gras."

In other words, Kobe beef might be best served as an appetizer, not a whole steak. Regardless of how it compares to Certified Angus Beef, chefs seem to agree that there's a place on the menu for both.

However, one trend that will never go away is aging the product to improve the palatability and tenderness. Whether the beef is wet or dry aged is another story. The East Coast tends to dry age.

"There's a lot of confusion in this country on aging of meat. Once you start heading west, there's not a lot of dry aging," Doherty says. "You definitely want to let meat age, because the natural enzymes break down the meat fibers to make it tender in a natural way. With dry aging, the best results are with highly marbled meat."

1996

Monthly sales record in August with 25.4 million pounds

Independent packers, such as Tarpoff Packing Company, market Certified Angus Beef brand Prime

Branded BUBBA burger patties available in frozen-food aisles

"Science Behind the Sizzle" program offers continuing education credits to chefs through the American Culinary Federation

First USDA value-based grid identifies packer premiums for qualifying cattle

Trademark Integrity Protection System (TIPS) expands account monitoring and auditing

Brett McKee's Oak Steakhouse, Charleston, is considered one of the country's best. His signature dishes include Beef Carpaccio, Steak Pizzaiola, and Cajun Blackened New York Strip, all featuring the Certified Angus Beef brand.

At the 2009 Cayman Culinary Society Food Show, Chef Davide Pugliese prepares ribeye cap and top sirloin in trendy recipes for the tourist season. Pugliese is the owner and executive chef of licensed Brandywine Bay in the British Virgin Islands.

At Red, the Steakhouse, in Miami, executive chef and partner Peter Vauthy (left) orchestrates one of the most exciting steak concepts in recent years. Its sister location in Cleveland is also licensed and led by executive chef and owner Jonathan Bennett (right).

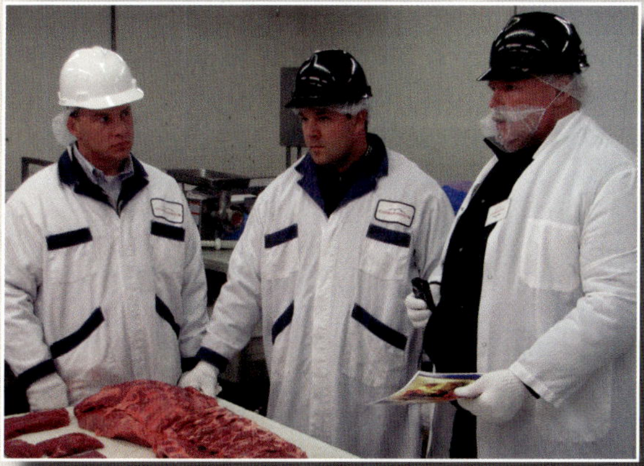

Offering the brand since 2000, the Cheesecake Factory's 150 locations nationwide showcase Certified Angus Beef burgers, beef ribs, Steak Diane, and Hibachi Steak. Thirteen Grand Lux locations also offer the brand. Staff member Clint Walenciak (left to right) visits with Donald Moore and Kix Nystrom-McGinnis, who lead the Cheesecake Factory's culinary team.

At the 2004 annual conference, Taste of Texas owner Edd Hendee relates how restaurateurs should not fear raising prices if the quality is there. It is a challenge he faced as well. Hendee's business, licensed for twenty-five years, is the brand's top volume independent restaurant in the world for steaks.

Buckhorn Grill does not mess around when letting people know just how good its tri-tip is. In one weekend, staff sample 780 pounds of the hand-seasoned and smoked beef at a California shopping mall. As 2009 draws to a close, owner John Pickerel plans eastward expansion of the concept — the Tri Tip Grill — in New York's Rockefeller Center.

J. Gilbert's Wood Fired Steaks and Seafood offers Certified Angus Beef Prime steaks. At its Kansas City area location in 2009, a Prime and Wine event shows off how its steaks are hand selected, expertly aged, cooked to perfection, and paired with leading wines.

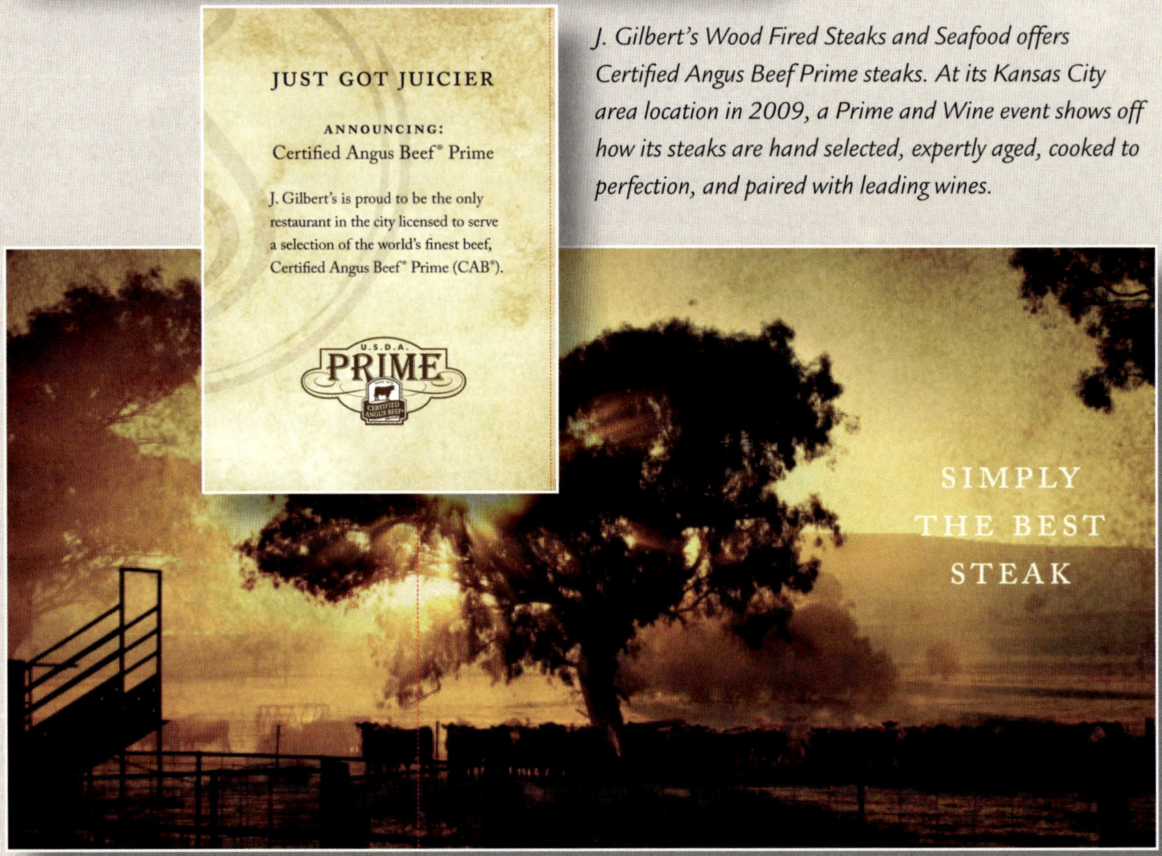

Greg's Steakhouse in Bermuda serves Certified Angus Beef products. The only true steakhouse on the island, it earns the brand's international marketing award in 2003. Accepting the award from president Jim Riemann (left) are owners Lorna and Greg Wilson.

Mexico presents a strong international component for some of the world's top eateries. Led by chef Esteban Wilson, the upscale Las Mañanitas resort restaurant in Cuernavaca is on the Condé Nast Traveler *2004 Gold List. In Mexico City, La Vid Argentina (below) features the brand at its upscale Argentinean steakhouses.*

Nueva Cocina Mexicana

Filete Certified Angus Beef en salsa de Huitlacoche y Tuétano

Ingredientes

1 880g. de filete Certified Angus Beef
2 160g. de ejote francés

Guarnición
- 160g. de calabaza italiana
3 4 piezas de souflé de elote
- 160g. de papa cambray
- 360ml. de salsa de huitlacoche y tuétano

Salsa de Huitlacoche y Tuétano
- Salsa demi glace
- 15g. de azúcar mascabado
- 22.5g. de tuétano finamente pi
4 60g. de huitlacoche molido
- 7.5ml. de vino tinto
- 30ml. de vinagre balsámico
5 80g. de cebolla picada
6 60g. de ajo picado

34 MéxicoGourmet

Most consumers spend less time in the kitchen these days—just as they prefer. As this trend develops in the 1980s and '90s, licensed meat companies begin making Certified Angus Beef frankfurters, "heat and serve" entrées, and sandwich meats.

9 Modern Convenience

While chefs seek the highest quality products to serve customers, they also look for ways to save time and money. But those customers, in their changing 1980s world, had already established strong demand for fast-food dining and quick-meal retail items. Women were entering the workforce in large numbers and time for cooking dinner wasn't a consideration until the evening. That's why people were cooking less and buying more frozen meals to pop in the microwave. From Uncle Ben's Rice to Hamburger Helper to Swanson's line of TV dinners, convenience foods were gaining popularity. But there was room for a high-quality beef line.

At the time, 80 percent of the program's sales came from the middle meats—rib, strip, and tenderloin—while the end meats from the chuck and round were not widely emphasized except in the international markets. And these were not sold in some of the traditional forms, such as roast beef, fresh corned beef, hamburger, hot dogs, and sausages. Festival Foods created the program's first summer sausages in 1986. Then, in 1989, Certified Angus Beef jumped into the hot dog world with an alternative for the convenience oriented customer.

Some questioned the move. After all, the program was based on premium standards and hot dogs were known for containing pork and poultry. Wouldn't a Certified Angus Beef frank diminish the name? Not necessarily. The brand's franks were made from chuck, and contained one-third less fat and sodium than the USDA limit.

"The bottom line on processed meats is their importance to the meats industry," wrote Roger Clark, former director of the brand's foodservice division in a 1990 *Angus Journal* article.

1997

Board approves 13,000-square-foot office expansion, with training room and professional kitchen

"The Sky's the Limit" banner from 1995 annual conference flies on Space Shuttle Discovery with Angus producer Colonel Mark Lee

Website launches (*certifiedangusbeef.com*)

First restaurant chain for deli: Mr. Subb has 33 New York locations

Licensing two additional Monfort (JBS Swift) locations and Moyer Packing brings access to 80 percent of nation's fed cattle

The brand's convenience items include frankfurters, burgers, precooked prime rib, fresh corned beef brisket, and more. Since forming the value-added products division in 1994, Certified Angus Beef has licensed many top-tier companies to grow this burgeoning market.

Much like the Natural and Prime extension lines many years later, Certified Angus Beef licensees showed the company there is consumer demand for high-quality, value-added products. In fact, processors sold such items for eight years before Certified Angus Beef created a value-added products division. In 1994, the company formed the division and committed to helping licensed participants merchandise the entire carcass, while providing quality-conscious consumers the convenience items needed for their busy lifestyles. It quickly blossomed.

According to company documents, the division placed emphasis on the frankfurter and deli meats. Besides the cornerstones of quality and product safety, the staff focused on service efforts like promotions, point-of-sale materials, and product development. As for supply, that would come from the anatomical facts: Most of the beef animal is comprised of end meats, but most of the Certified Angus Beef fresh product sales at the time were in middle meats. In addition to the frankfurters, hamburgers, and roast-beef-sandwich meat, processors were soon creating premium microwavable meals.

In 1997, Jon Rocke at RMH Foods, in Morton, Illinois, introduced several minimally processed Certified Angus Beef products, including Heat-N-Serve Beef Pot Roast with Gravy, Beef Sirloin Tri Tip Roast, and Prime Rib aus jus. The raw beef product was shipped directly to RMH from a packer. Employees hand-rubbed the beef with season-

When marketing something as quintessential to American families as home-cooked meals, it's hard to overcome the "grandma makes it better" notion. But that's exactly what RMH Foods does with the slogan: "Whatever You Do ... Don't Tell Grandma!" Company president Jon Rocke and grandmas are pictured in promotions.

ing and slowly cooked it. Unlike similar products on the market, these were not pumped full of fillers, water, and phosphates. The items were vacuum-packed and sent directly to retailers. They were a huge hit right away.

RMH Foods tested them at two Rocke's Meating Haus retail outlets. The test was so successful that a representative from the Sunflower Group Demo Agency, which aided the product rollout, wrote: "Never in all my test kitchen days have I seen samples disappear as quickly as yours did! ... You have a wonderful product that I expect will be very well received."

The customer reviews shared similar praise. One skeptical tester said: "Precooked, huh? I don't like precooked meats. Oh, but this is delicious!"

The biggest kudos came from the industry. In 1998, two of the products were named as top five finalists for the National Cattlemen's Beef Association (NCBA) centennial "Best New Beef Product in America."

Unfortunately, that would not carry over for very long. Once larger companies saw the success of RMH Foods' Heat-N-Serve line and entered the category, the premium microwavable meal became "a glorified TV dinner," says Brett Erickson, director of

The Advance Food Company team gears up for a product sampling at the brand's annual conference. Representatives (from left to right) Rhobb Walter, Pam Godlove, and Rob McLaughlin find unique ways to market the product to distributors, leading it to top restaurants.

One Advanced Marketing Company

Advance Food Company is one of the top food processors in the country, according to *Refrigerated & Frozen Foods Magazine*. The company has a strong standing in the restaurant industry and has been licensed with Certified Angus Beef since 2001.

In 2008, the brand and the processor partnered on ads in multiple restaurant industry publications, including *Restaurant Business* and *Restaurant Hospitality*. The Oklahoma-based company also hosted a booth at the National Restaurant Association show in Chicago, which has more than seventy thousand attendees annually from all fifty states and more than one hundred countries. "We were able to showcase eight Certified Angus Beef products with top restaurants," says Pam Godlove, Advance Food Company marketing coordinator. The elite eight were country-fried steaks, Philly steaks, corned beef, burgers, mini burgers, chili, brisket, and shredded beef.

Advance Food Company partners with the brand to offer several ready-to-cook products for foodservice, including country-fried steak. Made from sirloin and hand breaded, this product has received quite the following in restaurants nationwide.

In addition to winning over grandmas, RMH's Certified Angus Beef products win several industry awards, including the National Cattlemen's Beef Association's Best New Beef Product 2000. RMH Foods president Jon Rocke (left) and his wife, Jolene, accept the grand prize from Carl Blackwell, NCBA executive director of new product development.

value-added products. "RMH was trying to compete against the larger manufacturers that were offering a commodity product."

As the premium meal-ready entrée opportunities seemingly dried up, the value-added division opened new doors in other areas. In 1996, the brand licensed Bubba Burger to sell frozen patties in retail. Bubba Burger carries several lines and receives a lot of prime-time television exposure through its sponsorships of the Chicago Cubs, Philadelphia Phillies, Atlanta Braves, New York Mets, and Jacksonville Jaguars. These venues, combined with retailer promotions, create a loyal following for the brand's Bubba Burgers.

Holten Meat Company is another value-added licensee that has won over America's hamburger-eating audience. Known for its high-quality Thick-N-Juicy patties, Holten had become an industry leader for retail-ready burgers.

When Certified Angus Beef called, owner Jim Holten says licensing in 1991 was the perfect match for his family-owned company. The burgers combine the rich, full flavor of Certified Angus Beef chuck with the company's secret seasoning.

"They know they can hang their hat on our patties because we use the same type of raw product consistently and I have twelve people in direct quality control, completely responsible for consistency of my product," Holten says. "We test-cook every batch of meat we produce. And Certified Angus Beef has helped us in the restaurant industry. We've been good for each other."

1997

More than one million pounds marketed per working day

More than 2,000 restaurants and stores visited during year to ensure proper trademark use

Joint ad for restaurants appears in *Continental Airlines* magazine.

Bubba Burger is a brand synonymous with convenience and quality. When Bubba Burger introduces Certified Angus Beef burgers made with sirloin tips in 1997, the brand has a spot in the frozen-foods section for the first time.

Another great match for Certified Angus Beef has been Advance Food Company, of Enid, Oklahoma, which produces a fully-cooked Pub Steak Burger. Made of Certified Angus Beef chuck, the product is a solution for hospitals, hotels, military bases, golf clubs, and other foodservice-related venues looking to save on labor while serving a premium burger. Licensed since 2001, Advance Food Company has carved out an opportunistic niche with its line of products, which include country-fried steak, chili, and Steak-EZE Philly steak.

In a 2003 interview, Rhobb Walter, national product manager for the company, said one sales point has always been, "We can make it as good as yours, more consistent, and with a lot less labor involved. With Certified Angus Beef, there's more equity in that statement. It's known for quality with steak cuts. It's allowed us to market the very best available. That's what the industry is looking for—something that stands out. We've got the best product. You're not going to find anything else better."

Pam Godlove, Advance Food Company's marketing coordinator, says the Certified Angus Beef line has gradually increased in sales over the years. "Everything they do is top class and it's a fun program to be involved with, but also something we can be proud of. They always talk about the science behind the specifications. That means a lot to me as a processor."

St. Louis-based Schnuck Markets advertise Holten burgers and Usinger's franks, giving customers the secrets to perfect grilling.

Godlove believes the brand's logo commands consumer awareness. "I don't know if I would say as high as Coca-Cola, but I do think people definitely recognize it."

While Coke may have an edge in brand power, the not-for-profit company scored that level of press attention at the 2002 Olympic Winter Games in Salt Lake City, Utah. As the official beef supplier to the Games, the producer-owned company provided beef products for concessions. One of those, a Usinger's frankfurter, made with Certified Angus Beef chuck and the finest spices, was such a hit that the concessions nearly ran out. To avoid that, the company moved its supply of franks from the media tent to the paying eating establishments. The brand sold a record one hundred franks per minute at the Games.

When the media showed up for lunch, they were all asking about the hot dogs. "That became a story, that the Olympics were running out of hot dogs," says Fritz Usinger, of Milwaukee-based Usinger's. "The follow-up story was about this company in Milwaukee working overtime to get a truckload of hot dogs to the Olympics in a crisis situation. Throughout the Games, it became a little sidebar story about how Certified Angus Beef is meeting the needs to deliver hot dogs to the Olympics."

Those were certainly treasured memories for the Usingers, but the frank has been a hot seller wherever it's sold. In fact, so many non-licensed retailers have requested the product that Certified Angus Beef had to restructure its policy for consumer-packaged

The Real Frank Story

The 2002 Olympic Winter Games food committee, charged with making sure the concessions had enough product, forecast a need for one hundred thousand pounds of the Usinger's Certified Angus Beef frank. Brent Eichar remembers hearing that number and questioning how accurate their forecast was.

"Within about three days, we were running low," he says.

Eichar, who says he's not a risk taker—"I'm the finance guy"—called Usinger's and requested another forty-thousand-pound load of franks be produced and shipped to Salt Lake City that night. The ever-conservative Eichar admits, "I was taking a gamble here. It's a $100,000 decision."

Usinger's needed more chucks to fill the order so Eichar called Rob Shuey at Tyson Fresh Meats to have a truckload rerouted to Milwaukee. "Rob found a load going to a retail distributor and diverted them literally in the process of being shipped. Usinger's produced those franks, put them on an airplane, and shipped them to the Games."

Eichar later found out the product arrived just in time. "It was within an hour or two of them running out. So we got all of this publicity for the 'sold-out frank.' But we never technically ran out."

At the 2002 Olympic Winter Games in Salt Lake City, Usinger's Certified Angus Beef franks are a huge hit. For a Johnny Biscuit television report, staff form a frank pyramid reminiscent of the medal podium for the 100-franks-per-minute sold at the games.

items so all retailers could carry the frank or other value-added products. Usinger's offers several of the brand's products, including four types of franks, summer sausage, Polish sausage, and Andouille. Looking back to when the brand first approached his company, Usinger says it was the perfect opportunity.

"There was not much debate at all," he says. "To have your brand associated with their premium program was a no-brainer."

The 2002 Olympic Winter Games did not mark the only time value-added products received major airtime. Gary West Smoked Meats' Certified Angus Beef Steak Strips, licensed in 2004, have been featured on *Good Morning America*, the Food Network, CNN, and QVC, as well as *The New York Times* and *Outdoor Life*. Not to be confused with beef jerky, which is typically ground and formed into stick shapes, these steak strips come from inside rounds and are smoked the old-fashioned way over burning hardwood for natural flavor and steak-like texture. The strips are so tender, "they almost melt in your mouth," Erickson says.

Just as critics questioned the frankfurter, many asked, do we really want to be in the snack category? Unlike most snacks, these Steak Strips are not sold in gas stations. Instead, the product is primarily purchased through "outdoors gourmet" catalogs, like Cabela's, and some Las Vegas hotels include them in their rooms at the service bars. In addition, Gary West promotes the product at the National Finals Rodeo.

"People are looking for healthy and convenient, hand-held snacks and this fits that segment," Erickson says. In fact, the Mintel International Group, a consumer research firm, reported meat snack sales grew 147 percent from 1997 to 2002. Gary West Smoked Meats, Jacksonville, Oregon, saw that growth trend and the premium beef brand as a great fit for a product that can trace its recipe to one the owner's family used as their wagon rolled along the Oregon Trail.

"Marbling makes these steak strips more moist than typical beef snacks," said Paul Murdoch, owner of Gary West, in a 2004 interview. "The Certified Angus Beef brand gives us a distinctive edge."

It also gives King's Command Foods a distinctive edge in the meatloaf category. The company based in Kent, Washington, has been selling rotisserie meatloaf to retailers since 2004, placing it next to the chicken in the hot deli section. Initially a restaurant meal solution, the meatloaf became such a hit that the company began providing it to just about every major retailer.

Once it had proven the market was strong, Certified Angus Beef approached King's Command in 2007.

"There are certain Certified Angus Beef stores that have a clientele used to buying very high-end product, and this meatloaf actually gives them the opportunity to serve a cooked-in-store version," says Ron Baer, president of King's Command. "A lot of times people don't have an hour or the knowledge of how to cook meatloaf, so having a restaurant-quality product, cooked in a retail environment with retail margins is big."

The product is made with chuck and the freshest ingredients, so it's like homemade. King's Command packages several frozen meatloaf items with the sauce and line-by-line instructions for the retailer to follow.

"We have tight controls from the raw material, tempering of the materials, to the finished grind, to the blend time," Baer says, adding that the brand has been an ideal partner.

"Certified Angus Beef is a great, quality-assured product. It has a restaurant following and a grocery store following," he says. "The consistency is important on their side and they track it very well. As far as merchandising, they're one of the finest."

Kober, the brand's retail director, believes the product will be a hit. But there are so many family recipes out there that not everyone will find it is "how Mom used to make it." Still, he and Erickson say the timing is just right for the meatloaf category.

"The hot deli area is perfect right now because more consumers are leaving the restaurants and going into stores looking for the convenience factor," Kober says. "You don't see much branding or promoting in the hot deli section. For retailers, the meatloaf is an opportunity to differentiate by carrying a high-quality line of hot foods."

1998

Beef Brief reports 9.14 million cattle evaluated during 1997 for USDA certified brands; of all beef certified, nearly 90 percent sold as Certified Angus Beef brand

First menu designed by staff: Charlie's Chop House

Consumer recipe book marks brand's 20th anniversary

Increased emphasis on supply development leads to new divisions for industry information and feedlot licensing

First licensed feedlot: Triangle H Grain & Cattle Company

Left: The brand's meatloaf from King's Command Foods has been sold next to rotisserie chickens in the hot deli since 2008. The product gives shoppers a beefy alternative to ready-made chicken.

Below: In 2004, the brand partners with Gary West Smoked Meats to offer steak strips—not to be confused with beef jerky. Made from inside rounds and smoked over hardwood, this hand-held snack is found in stores as well as gourmet catalogs, outdoor stores, fine resorts, and television shopping networks.

1998

Consumer marketing focus begins in Houston and Jacksonville with "Life Can't Have Too Much Flavor" campaign

Retail middle-meat marketing program results in 49 percent increase in same-item sales over first four months in 1997

First cross-promotion with another brand: Borden Soup Starter

Chili's Bar & Grill licenses for country-fried steak at 500 restaurants

Pot roast and prime rib from RMH Foods (formerly Rocke's Meating Haus) named Best New Beef Products by NCBA

The Deli Dilemma

In most grocery stores, the delicatessen or "deli" manager does not work with the fresh meat department. "They're completely different divisions," Kober says.

When Certified Angus Beef account managers sell the roast beef, they have to go to the deli director. "Even if we have a great relationship with the meat manager, there is little carryover to the deli department," Kober says. And the competition in deli is much different than in fresh meat. Instead of competing against solely roast-beef makers, Certified Angus Beef takes on the likes of Boar's Head and Dietz & Watson, which provide turkey, ham, roast beef, salami, pepperoni, pickles, mustards, cheeses, and more. Retailers often choose one deli processor to keep administrative costs down and to keep things simple.

That has posed many challenges for the value-added products staff. But because the brand's roast beef has won so many blind taste tests with its superior quality, they're continuously searching for ways to make it work.

"Consumers have no idea that most deli meats are pumped full of binders, fillers, and water. In some beef products, it can be 30 to 40 percent water. With Certified Angus Beef, we allow a maximum 10-percent seasoning solution and will cook the product back to its natural weight," Erickson says.

But the retail market is driven by gross profit, not necessarily quality. There's more margin to be had with a 30-percent (mostly water) pumped product if you can sell it at a beef price, so the consumer often pays a higher

Sacramento-based Raley's stores introduce deli meats in spring 1997.

Opposite: Floor graphics lead Meijer shoppers to the deli for Certified Angus Beef roast beef. Jointly marketed with the Markets of Meijer and Certified Angus Beef brands, the product is sliced to order.

price for lower quality. "When Certified Angus Beef is partnered with a premium retail brand that offers a full line of deli meats, it increases our brand position within the deli while elevating the quality image of their entire deli line," Erickson says.

The Certified Angus Beef deli roast beef is available in at least eight major retail chains across the U.S., and does well in the H-E-B stores in Mexico. But perhaps the product's crowning achievement was the licensing of Mr. Subb in New York in 1997 for deli corned beef. Another is the Daily Grill, with corned beef in its twenty-two California and metropolitan locations.

Chef Scott Popovic (right) interviews with and cooks for WEWS-TV, the Cleveland ABC affiliate. Public relations and demonstrations are crucial tactics in Certified Angus Beef's marketing strategies.

10 Power of Marketing

When members of the Salt Lake Organizing Committee for the Olympic Winter Games called Certified Angus Beef in the late 1990s, they were searching for a food sponsor. The leading beef brand only had to come up with $3 million to become one of ten food suppliers at the 2002 Olympic Winter Games in Salt Lake City.

"That was still a huge number for us," says Brent Eichar, senior vice president of the company.

Owning no product, the staff approached packers about the opportunity. A rebate program then in place offered an option; the major packers agreed to shift those dollars back until after the Games. Packers and processors agreed to provide product, and by the time 2002 came around, the company had completely paid off the suppliership fee. Eichar considers the payment "a big deal for us to come up with the resources as a fairly small company."

As soon as the deal was announced, the company began planning. But nobody was solely dedicated to the event. In January 2000, a young Deanna Walenciak submitted a letter to the executive staff. "It outlined what a job description would be for the Olympic coordinator and what it would look like to have somebody focus on that job full time," she says. The team asked Walenciak if she would take on that job.

She agreed to it.

As the company's Olympic coordinator, Walenciak put her arms around everything. One of her first projects was working through the details of how to use the Olympic supplier logo, not the Olympic rings that world sponsors can use.

In February 2002, the Olympic Winter Games in Salt Lake City hold the attention of the world. While watching athletes from seventy-eight countries compete for gold, the Games' attendees, media, and athletes enjoy Certified Angus Beef items, including the fan-favorite frankfurter.

"There were so many rules and regulations around using those Olympic marks, and everything had to go through such an approval process," she says. "For example, you can't say the Olympics endorse your product; you're supplying your product to the Olympic Games."

The coordinator would spend many hours with the Games planning committee, while the Olympics' chefs created foodservice programs with Certified Angus Beef products. After they created a recipe for the brand's chili, Advance Food Company and Certified Angus Beef took that recipe and produced it for the Winter Games.

Even before the Olympics, the culinary staff fell in love with the frankfurter. It won a taste test against the nation's leading brand. The results were featured in an August 2000 *USA Today* story.

"We literally searched to find the best hot dog in the U.S.," said Don Pritchard, director of Foodservices for the Games, in the newspaper account.

At Olympic training centers around the country, Certified Angus Beef's frankfurters, barbecue beef, pot roast, and three deli meats were served. That exposure generated buzz among the committee.

"We assure the best possible products will be served to athletes, officials, spectators and the media by joining forces with Certified Angus Beef," said Mitt Romney, chief executive officer of the Games, in the story in *USA Today*, which would later cover his run for the Republican nomination in the 2008 presidential election.

Even before the first game was played, Certified Angus Beef was winning quality exposure. The question was: Did any of the wonderful attention trickle down to licensees? Worried that it wasn't happening, Walenciak focused on incorporating marketing programs for retailers, distributors, and restaurants. For licensed meat departments, the company created point-of-sale materials connecting Certified Angus Beef to the Olympics. One favorite was a nylon flag with a photo of Olympic speed skaters. The company also worked with processors to create new packages with the Olympic supplier logo for the featured value-added products. And for those who wanted to commemorate the event, a brass cowbell with an etching of the Certified Angus Beef and

Left: Sales contests and a consumer sweepstakes build excitement for the Games and featured products. Meat manager Dick Teutschmann (left), from Dick's Supermarket in Monroe, Wisconsin, wins tickets for himself and his wife, Lauri. He is pictured with Rocky Honeywell of Certified Angus Beef.

Above: Staff member Jason Clever is nominated by coworkers to carry the Olympic Torch in Omaha, Nebraska. For sixty-five days, more than 11,500 people participate in the time-honored tradition of relaying the flame to the Games.

Salt Lake Games logos was made available. In August 2001, the company launched a sweepstakes for consumers to "Win an Olympic Trip." Promoted at retail meat cases, the countertop display raved about Certified Angus Beef being a proud supplier of the 2002 Games.

Walenciak says these were great tactics, but the real payoff was everybody's enthusiasm and excitement for the Games. "There was just so much energy," she says. "People at those companies whose products were going to be served at the Games—Usinger's being the biggest with their frankfurter—were really excited."

Everybody wanted to go. The company's marketing team put together packages for licensees to attend the Games for free. Licensee executives received tickets, and the program held various contests for licensee employees.

At Certified Angus Beef, employees had to apply. When one staff member learned she had been accepted to attend, she was ecstatic. "She'd never traveled west of the Mississippi," Walenciak says.

Retired Certified Angus Beef president Jim Riemann says it was an opportunity for the organization to thank licensees that had worked hard at promoting the brand.

"It was so amazing that almost all of these people would come together initially as strangers, but by that last night they were there, it was almost like a high school graduation party—they didn't want to leave," Riemann says.

Longtime Loyalty

Some brands come and some brands go, but the Certified Angus Beef brand has staying power. Dozens of businesses have stuck with the brand for more than twenty years and continue to enjoy offering this premium beef to their customers and consumers. Key factors for their longevity include marketing and education.

EXPORTERS
Day-Lee Foods, Sante Fe Springs, California
MacDonald Meat Company, Seattle, Washington
Oritz International, Burlingame, California

FOODSERVICE DISTRIBUTORS
Cusack Wholesale Meats, Oklahoma City, Oklahoma
DeBragga and Spitler, New York
Dee's Foodservice, Albuquerque
Dole & Bailey, Woburn, Massachusetts
Facciola Meat Company, Fremont, California
George Wells Meat Company, Philadelphia
HFM Foodservice, Honolulu
Lombardi Brothers Meats, Denver, Colorado
Lone Star Foodservice, Austin, Texas
Malcolm Meats Company, Northwood, Ohio
Maryland Quality Meats, Baltimore
Neesvig's, Windsor, Wisconsin
Newport Meat Company, Irvine, California
Performance Foodservice Middendorf, Saint Louis, Missouri
Triple J Five Star Wholesale Foods (Pacific World Enterprises), Guam
Provimi de Puerto Rico, Vega Baja, Puerto Rico
Rocky Mountain Gourmet Steaks (Imperial Meats), Missoula, Montana
Scavuzzo's, Kansas City, Missouri
Sierra Meat Company, Reno, Nevada
Southern Foods, Greensboro, North Carolina
U.S. Foodservice/Stock Yards Meat Packing Company (City Meats & Provisions), Phoenix

PACKING COMPANIES
Cargill Meat Solutions (Excel Corporation), Wichita, Kansas; Schuyler, Nebraska; and Fort Morgan, Colorado
National Beef Packing Company Liberal, Kansas; Kansas City, Missouri; and Dodge City, Kansas
PM Beef (Caldwell Packing Company), Windom, Minnesota

RETAIL DISTRIBUTORS
Colorado Boxed Beef, Auburndale, Florida
Oxford Trading Company, Westwood, Massachusetts

SPECIALTY PRODUCTS
Hudson Meat Company, Columbus, Ohio
RMH Foods (Rocke's Meating Haus), Morton, Illinois

An ice-carving tour with Chuck "The Iceman" Cooper serves as the top venue for leveraging Olympic spirit with licensees, consumers, and media. Nineteen events capture more than twenty million consumer impressions and this Today Show *appearance.*

He joined the other employees to scrub pots and pans, delivered product to serving stations, hosted receptions, and stayed up late; "but it was amazing," he says. One licensee pulled him aside and said, "Jim, when your team came in and talked to us about participating in this suppliership, we had great doubts. But when I see how hard all of you are working out here, how people are experiencing the Certified Angus Beef brand frankfurters and chili, I'm convinced you made the right decision."

Then, there was the megaphone guy who steered people to the medal ceremonies. When the Certified Angus Beef crew and the megaphone man crossed paths, he blared through his megaphone, " 'Make way for the Certified Angus Beef people, they're the ones with those great frankfurters.' And I said, 'Man, we have arrived,' " Riemann says.

Attendees at the Games fell in love with the frankfurters, as evidenced by the two-hundred-plus stories about the "sold-out franks." The story was updated throughout the general Olympics coverage. In one BBC story, the columnist wrote about how the limited supplies "caused such a huge problem that organizers have agreed spectators can bring in their own food—including hot dogs—until new provisions arrive." As Eichar pointed out in Chapter 9, they never really ran out. And the Certified Angus Beef pins, which people would trade on eBay and in curio shops, have become an Olympic collector's item.

"Creating that kind of brand awareness is what it's all about," Riemann says. In addition to the $3 million suppliership fee, the hospitality program budget was $500,000, of which $90,000 covered the tickets. But the program was worth every dollar, officials say. For Riemann, the event would not have been possible without the help of licensees, Walenciak, and such a devoted staff. But for the brand, he says, "The real testament is when the head chef came to us and said, 'We know those people attending the Olympics

1999

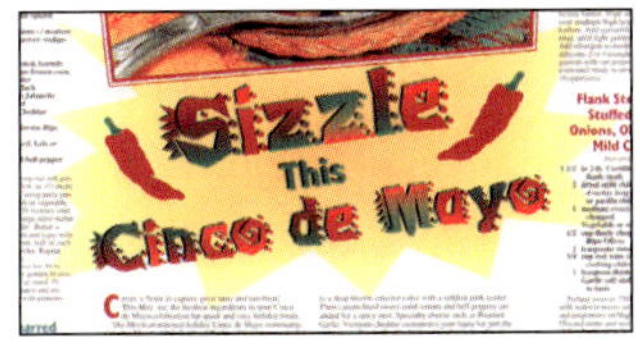

First full-color recipe page for newspapers celebrates Cinco de Mayo

First quality assurance seminar held for licensed feedlots

Weber Grill Company features brand in *Grill Out Times* newsletter, reaching 450,000 homes nationwide

Line-item licensing gives restaurants more flexibility in menu offerings

"Certified Clyde" becomes brand's modern mascot

"Here's the Premium" long-term study begins reporting premiums for Angus over non-Angus calves at U.S. auction markets

Marketing materials, pictured for 1993 (right) and 2007 (above), lead customers to the brand in stores and restaurants.

are going to experience top-quality athletic competition, but we want them to also have a top-quality food experience while they are in Salt Lake. And we want Certified Angus Beef so that we can deliver that top-quality food experience.' "

Research after the Games indicated the suppliership made great business sense, too. According to the Games Sponsorship Overview, more than 80 percent of the athletes agreed that sponsorships contribute greatly to successful Games. And in an eleven-country study, 68 percent of consumers favored commercial associations with the Olympics. According to media analysis statistics, consumers saw or heard the Certified Angus Beef brand mentioned nearly twenty-seven million times on television, more than a half million times on radio, and fourteen million times in print coverage. As Walenciak hoped, the coverage had an awesome trickle-down effect for licensees. At the end of the Games, Advance Food Company's Tim McLaughlin posed a familiar question: "Could this get any better?"

For Walenciak, the Olympics were much more than just numbers. It boosted employee morale, built brand awareness, and more. "It gave people a new reason to talk about Certified Angus Beef."

Target Markets

In the 1990s, the company made a major shift in its marketing programs. Instead of focusing on business-to-business promotions, the organization would now target consumers. The new philosophy would not necessarily ignore the agriculture, meat, restaurant, and retail trade journals, which regularly wrote about Certified Angus Beef. But there was a genuine shift in marketing to consumers.

The idea was to go from a "push-to" strategy to more of a pull-through. In some respects, that was how the brand was already being marketed. But the company wanted more emphasis on the consumer to become a "destination product," says Tracey Erickson.

To kick off the new consumer focus, a hot air balloon was launched from the Certified Angus Beef parking lot. The local paper extensively wrote about it. "Really, that was our visible representation that we were transitioning to things more newsworthy, with more public-relations coverage," Erickson says.

They then looked at how to create a national newsworthy marketing scheme. One of the first notions was to create a vehicle and drive around the country to create brand awareness like the Oscar Mayer Wienermobile. Another idea was to become a NASCAR sponsor and have the logo on a car hood. They also looked at signing a big-name national spokesperson like Martha Stewart. But these ideas were proven fiscally impossible and there were questions on whether a large-scale national campaign would help local licensees.

"We needed to establish some stronger ties in the communities," Erickson says. "We established more of a local focus."

In 1998, the brand launched its first consumer campaign, "Life Can't Have Too Much Flavor." While Certified Angus Beef had targeted consumers before with general advertisements, this campaign's messages were strategically connected through radio, billboard, and print. The underlying theme was bringing "the very best-tasting beef into the consumer dining experience whether they're eating out or buying products at the retail store," Erickson says.

Throughout the 1980s and '90s, taglines deliver a cohesive message in stores. They include "The Only Way to Be Sure," "Taste the Difference," and "Clearly a Matter of Taste."

Instead of plastering those messages all over the country, the brand focused on Houston and on Jacksonville, Florida. In its first year, the campaign generated ninety-two million impressions through fifty-two hundred metro spots, four thousand traffic radio sponsorships, one hundred and fifty features in retailers' advertising circulars, and regional newspaper and magazine advertising. The goal was to boost licensee sales in those areas and give everybody the same message. "For the first time, we were able to bring our licensees something with a cohesive message," Erickson says.

The Houston and Jacksonville experiment worked, and the tactic would become known as "Target Markets," the company's most important marketing initiative. In addition to local ads, licensees picked up great press. Added to the mix in 1999 were Boston and St. Louis, as well as secondary markets in North Carolina, Michigan, Wisconsin and California.

Over the course of four years, Certified Angus Beef pumped millions of dollars into such target markets in an attempt to drive consumers to licensed operations. A 1999 *Arlington Morning News* article touted Houston-based Saltgrass Steak House as one of the best steakhouses in Texas with the headline, "Saltgrass lets steak speak for itself." Buck-

In 1998, as the company directs more marketing efforts to consumers, it also develops more comprehensive, creative advertising campaigns. These promotional tools are offered to all licensees, and the brand places these ads in target cities.

1998

2000

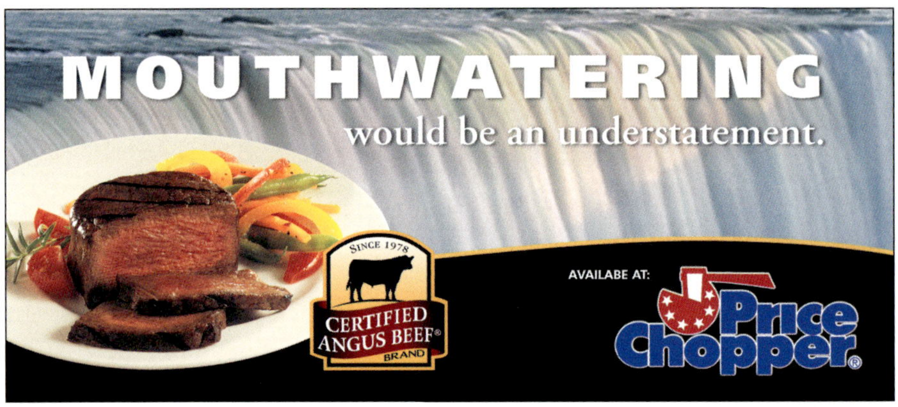

2003

2007

Staff chef Michael Ollier leads a cooking demonstration at the Greater Cincinnati Kitchen, Bath, and Remodeling Show. He discusses beef cuts, cooking techniques, and new recipes with consumers. Cincinnati is a 2009-2010 target market for the brand.

Target Market Evolution

Prior to the target market initiative, Certified Angus Beef considered using consumer marketing dollars on national advertising.

But an Australian consultant, Alf Nucifora, came in and said the group didn't have the budget to effectively reach all consumers at that level. He said money was better spent on event marketing and localized public relations.

As they became more experienced with target markets, staff focused on demographics first, doing their homework before initiating tactics. They met with licensees in the area and figured out the best way to target consumers through print, radio, or events. "Opportunities were out there," Walenciak says. And those public relations and event strategies have been most effective.

Grassroots Marketing

Certified Angus Beef is a company founded on grassroots efforts. That transfers over to marketing.

In the 1980s, the company started handing out bumper stickers that said "Any way you slice it, it's America's finest beef." In the 1990s, the brand made its way in target marketing. And today, it's reaching the world via the Internet.

On *YouTube.com*, corporate chef Popovic is somewhat of a celebrity. Certified Angus Beef has its own channel on the popular website, where Popovic gives demonstrations about how to pan roast, sauté, grill, season, and other cooking techniques. All videos start out with "expert tips you can trust" and the average video garners about six thousand viewers.

The flagship consumer-facing Internet marketing campaign is based at *certifiedangusbeef.com*, where viewers find everything from beef cuts and quality, to cooking and grilling tips. Consumers can also download recipes and rate them, after checking out their nearest licensed restaurant or retailer on the "where to buy" page. Boxes of steaks for gifts are an option as well.

Left, top: A cooking with kids contest launches at a 2007 home and garden show. Devin Whitney (right) is congratulated by Jerry Wallace (left) from Price Chopper, as brand chef Scott Popovic registers his approval.

Left, below: With its consumer-focused marketing strategy, the company supports licensed partners with billboard, radio, print, and Web advertising. The public relations team lands radio and television interviews, and holds product demonstrations at local home and garden shows.

Above: Fun activities include shopping-cart races and grill giveaways at Major- and Minor-League baseball games.

horn Steak & Roadhouse had a Knight-Ridder story published in newspapers all over the country in early 1999. From the story: "The steakhouse owners say their claim to fame is serving Certified Angus Beef with a motto, 'Real food for people who know the difference.' "

While the press coverage seemed to lean slightly toward the restaurants, the retail meat departments greatly appreciated the extra focus. Through the target market initiative, Certified Angus Beef buys print and radio ads focusing only on the meat. To make sure they're hitting the media outlets, marketing staff at the brand work closely with the licensee's marketing staff.

Above: Aligning promotions with other great brands creates additional opportunities for reaching consumers. Louis M. Martini, Lea & Perrins, and other strategic partners cooperate on promotions to match brands with a similar focus on quality.

Opposite: This full-page advertisement appears in several issues of the Robb Report *for the Waldorf=Astoria's Bull and Bear. The iconic New York City restaurant wants the world to know it serves Certified Angus Beef Prime.*

"We absolutely learn from their marketing expertise," Walenciak says. "They know what type of events would work in their markets. At the same time, they have learned from us because we approach things very differently. We have smaller budgets and what we've found in retail is they have larger budgets to spend on big things, but they don't have a lot of staff to execute grassroots promotions."

As an example in St. Louis, licensee Schnuck Markets had about ninety stores in that area and a sponsorship with the St. Louis Cardinals. When the Certified Angus Beef team met with the company's marketing staff, they aimed to leverage Schnucks' sponsorship. The recommendation was a product sampling at a game and giving out coupons during the Seventh Inning Stretch. The focus was to drive people to a Schnuck Market after the game.

For this and other efforts, including its "meat manager school," Schnucks has consistently been one of Certified Angus Beef's top five retail outlets. The retailer has received a great deal of media publicity, including spots on local television networks. The retailer's proximity also helped Certified Angus Beef get its foot in the door with another major St. Louis brand. In 2007, at Schnucks, "Toast the Season" advertisements featured the Certified Angus Beef logo with a neatly cut roast and a tall glass of Michelob Lager. The promotion offered consumers \$3 savings when they purchased any Certified Angus Beef product.

The continuing goal of target markets is to reach women, ages thirty-five to fifty-five, with children, and an annual household income of \$50,000 or more. Men in the same demographic are a secondary target. Through direct-mail sheets, coupons, outdoor billboards, and inserts for newspapers and lifestyle magazines, Certified Angus Beef reached 93.8 percent of the primary St. Louis audience in 2003. Media research indicates these consumers actually saw or heard the marketing message 16.9 times.

That same year, corporate chef Dianna Stoffer traveled to the Dallas/Fort Worth Home and Garden show. She worked on the Gourmet Cooking Stage on the busiest day of the three-day event, which attracted some 45,000 consumers. In 2008, chef Popovic conducted cooking classes for Albany, New York, licensee Price Chopper. The local newspaper ran a short blurb on his demonstration and hundreds showed up.

"Along with cooking demonstrations, whenever we're in town we try to make extra media connections," Popovic says. "In Albany, we were tied into sponsoring World Team Tennis and we were able to be on radio and TV to talk about the brand."

Walenciak says the target markets are successful because of the partnership with licensees. After all, they know the local markets and customers the best. The 2009 and 2010 target markets include Cincinnati and Seattle, where Certified Angus Beef has been working diligently with the Meijer and QFC marketing teams to make a splash.

"They're thrilled that we're going to focus on their meat departments," Walenciak says. "Restaurants are equally excited to participate in our activities."

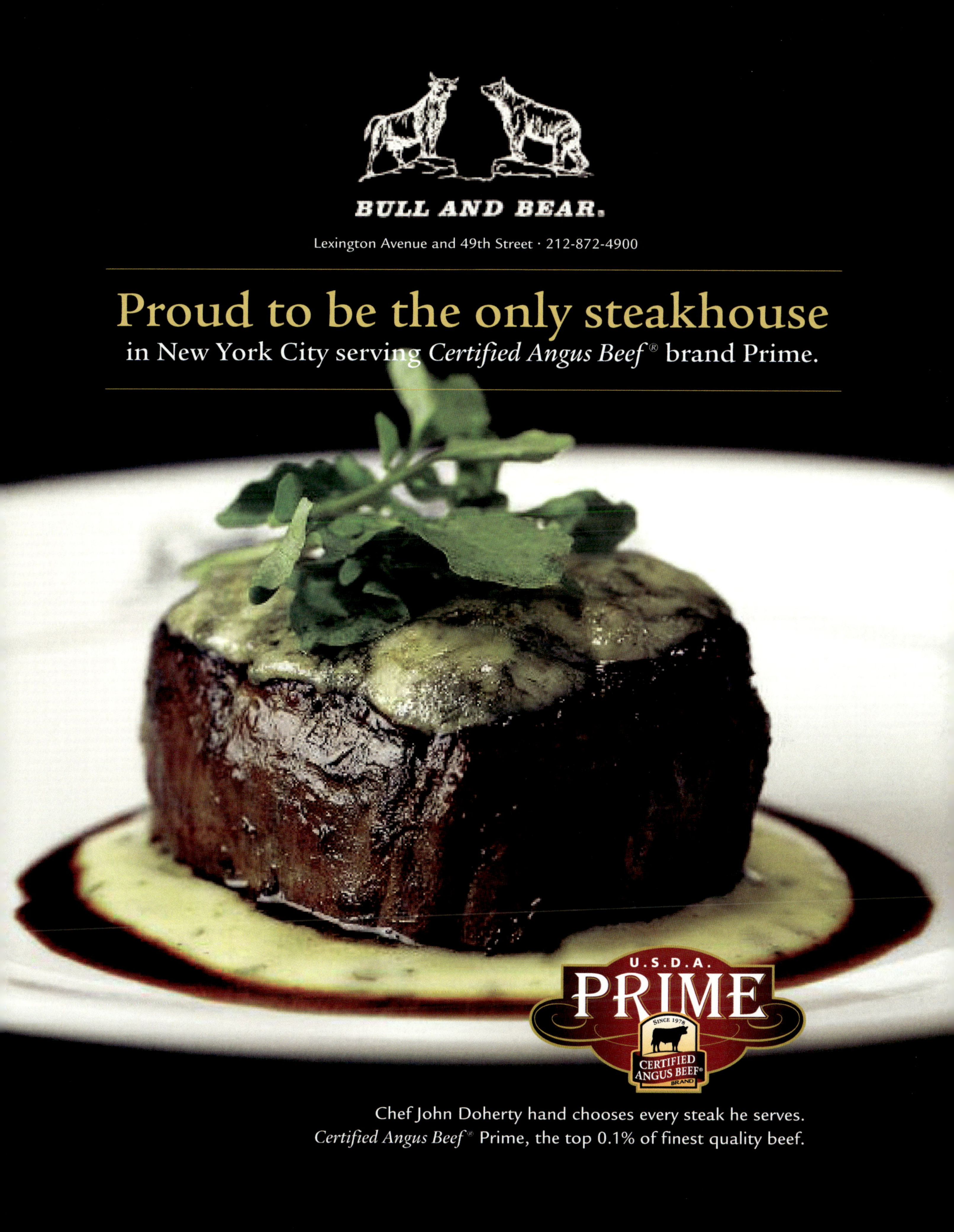

BULL AND BEAR.
Lexington Avenue and 49th Street · 212-872-4900
Proud to be the only steakhouse
in New York City serving Certified Angus Beef® brand Prime.
U.S.D.A.
PRIME
SINCE 1978
CERTIFIED ANGUS BEEF®
BRAND
Chef John Doherty hand chooses every steak he serves.
Certified Angus Beef® Prime, the top 0.1% of finest quality beef.

1999

First promotions held in Dominican Republic, Barbados, and Costa Rica; first seminars for chefs held in Peru and China

More than 3,500 menus collected and 1,000 stores visited this year

First direct upload of sales data paves way for modern-age product monitoring

Dr. Jim Riemann named executive director upon Mick Colvin's retirement

Via Christi Regional Medical Center among several healthcare facilities licensed

Reaching Consumers with Licensees

Cooperative marketing has long been considered one of the most effective and underused forms of reaching consumers. Two companies vying for the same consumer with different products execute advertisements together. The strategy is especially commonplace in the retail food world, where cooperative branding is common. When the shopper's eyes see literally thousands of brand names, manufacturers want them to focus on their logo in as many places as possible. Brands like Nestlé, Pillsbury, Healthy Choice, and Kellogg's can be found in almost every aisle as a partner with another brand. For example, Healthy Choice and Kellogg's make a cereal together and Pillsbury sells refrigerated dough made with Nestlé chocolate chips. These strategic partnerships are a win-win for all companies involved. And the cooperation makes sense, especially when linking complementary food brands of similar quality.

That's the strategy Certified Angus Beef employed when looking at possible partners. "Our goal is to marry up with brands or companies with at least an equivalent position to ours," Erickson says. "Hopefully, they have an elevated position where we could ride a bit on their coattails." The marketing plan could be especially useful for a not-for-profit company with limited funds.

Some obvious strategic partners have been companies associated with grilling. One ad featured a Labatt Blue beer and Certified Angus Beef steak that said, "Pick up Your Free Summer Grilling Guide at Meijer." In 1999, Weber Grill promoted product in its *Grill Out Times* newsletter, which reached 450,000 homes. According to the *Angus Journal*, half of the newsletter's readers learned about the filet mignon, London broil, Kansas City strip, and New York strip through the promotion.

E & J Gallo wine has been an extremely successful partner over the years. In 2002, the two companies teamed up to offer $2 coupons for Certified Angus Beef. "Our programs have been successful because our brands convey a similar image," said Tom Bernth, E & J Gallo Winery's trade marketing manager for the Midwest and central regions, in a 2002 interview with *Agri Marketing*. "The synergies between our brands allow us to be very creative in our advertising, POS, and execution."

In 2008, Certified Angus Beef announced a partnership with Lea & Perrins, to which many steak lovers would say, "What, a great steak doesn't need Worcestershire sauce." But Erickson says the brand has been very selective in choosing partners. Her team has turned down numerous steak sauce and marinade partnership proposals. "Lea & Perrins potently impacts the flavor of the product, but the key is in how we are using it," she explains. Target promotions are geared to burgers and cuts where a marinade or creative recipe makes sense. "Lea & Perrin's Worcestershire is aged for eighteen months and it's the premium product in the category."

In addition to partnering with other food product makers, Certified Angus Beef achieves a good deal of media reach through supporting licensed retailer and restaurant marketing efforts. In 2007 and 2008, licensees have been paired with Certified Angus Beef-paid ads in major national magazines, including Price Chopper's full pages in *Cooking Light* and *This Old House*; and Waldorf=Astoria's full page ad in the *Robb Report*. Gristede's Supermarkets, the first to execute a large national advertisement for the

The on-staff design team is the gatekeeper of "the look." Colors, photos, the wave, tagline, and logo are strategically and consistently placed to portray the brand's premium image.

Outsourcing Work

For the "Life Can't Have Too Much Flavor" and other campaigns in 1998, Certified Angus Beef outsourced $300,000 worth of design and copywriting work. And each time the design firms filed something, Erickson had to correct or fine-tune the brochures or advertisements. It didn't take her long to realize these advertising and design agencies were not providing the service or quality she wanted.

So she did something her mentor, Colvin, would have done: hired the right people, graphic designers who would understand the brand while growing in their creative abilities, to include food photography.

By using internal staff instead of outsourcing creative jobs, Erickson estimates the brand has saved about $5 million in ten years. And more importantly, the consistency, quality, and responsiveness have improved.

"Our folks can crank out a file much more quickly than if we were going through a third party," she says. "That's what ensures consistency from promotion to promotion. The creative team really does a great job."

In 2009, the team included four full-time designers who complete approximately seven hundred projects per year.

Licensed partners incorporate the brand in frequent promotions. This Food City "Endless Grilling Season" promotion and its sponsored NASCAR events in Bristol, Tennessee, include the brand.

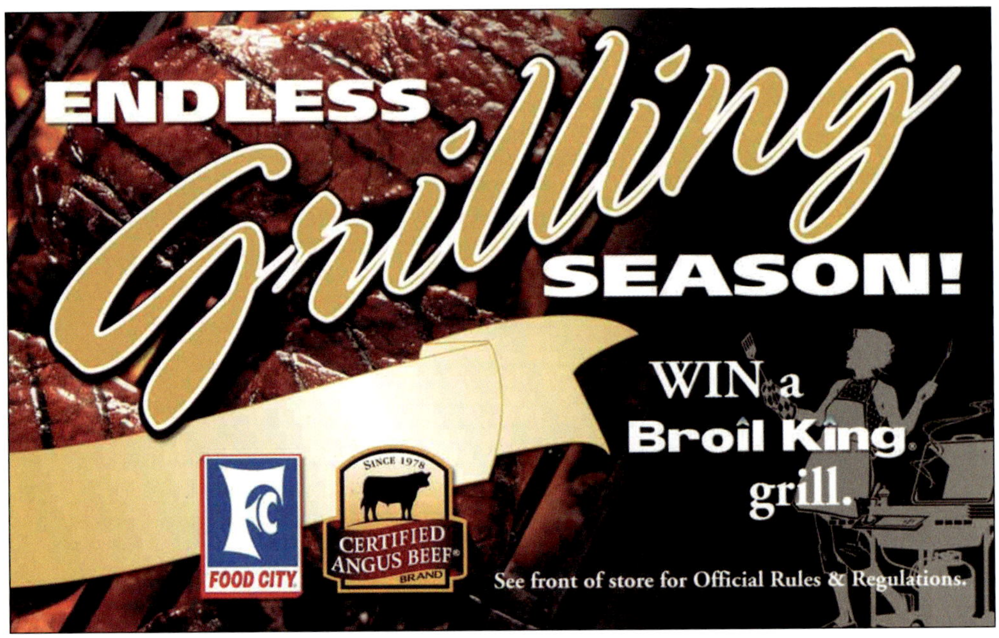

company in 1984, regularly advertises Certified Angus Beef in *The New York Times* and *The Wall Street Journal.*

Just like at the distribution level, licensed restaurants and retailers incorporate the brand into their own creative marketing efforts. In 2004, Food City hosted the "Endless Grilling Season" promotion and bought numerous radio commercials strategically designed to increase middle-meat sales. Two years later, Food City executed in-store cooking demonstrations and product samplings at NASCAR Race Nights. As Tony Stewart and the Home Depot car were on their way to win the NASCAR Nextel Cup Series, Food City's customers were patronizing the store even more. The retailer was

In May 2007, Donald Trump launches Trump Steaks, a premium line of Certified Angus Beef Prime steaks and burgers, with specialty retailer The Sharper Image. He appears on QVC with host David Venable (center) and staff member Sarah Donohoe. Trump's Atlantic City casinos serve the brand.

reaching them through television, billboard, print, and online advertising, while its well-trained staff was telling customers just how good Certified Angus Beef is.

To promote the brand in 2008, Trig's Food and Drugs of Minocqua, Wisconsin, featured radio segments on local stations, including live radio events at stores. The brand was prominently featured in the weekly circular and on a roadside billboard. The store's own local meat department celebrity, TomKat, sampled product and interacted with consumers to promote the brand.

That same year, Meijer introduced Certified Angus Beef deli roast beef. To drive sales and build awareness, the company executed its Markets of Meijer promotions and product sampling at the International Women's Show. Using the same kind of tactic, Buckhorn Grill, of Winters, California, touted its "Best Tri-Tip on the Planet" and sampled 780 pounds of Certified Angus Beef tri-tip roast at a local shopping mall.

For television in 2008, Famous Dave's used humor to promote its limited-time Smokehouse Steak. The commercial opens with a shot of a doorknob as the door slightly moves aside. It pans to a young couple sleeping. And then a black hoof enters through the door and two well-fed Angus steers rush toward the bed and flip it over. The woman screams as the couple falls to the ground and a deep, robust voice says: "People Tipping: Now that's Angus with attitude. Like Famous Dave's New Smokehouse Steak. Made of 100-percent Certified Angus Beef flank steak." Famous Dave's reported that the television spot "exceeded expectations."

Erickson says that's the kind of marketing that helps Certified Angus Beef prosper.

"We are absolutely dependent on the partnerships we have with our licensees, their commitment to the brand," she says.

And the winner is...

Licensed businesses are honored each year for their leadership in sales, marketing, and commitment to brand integrity. The awards are highly coveted, with many challenges between businesses for a fight to the finish. Volume leaders receive a pewter bull statuette, which was created by artist Jim Reno in 1983 to celebrate the Association's centennial anniversary and reproduced from the original by R.S. Owens, maker of the Oscars for the Academy Awards. Some businesses have a dozen or more in their "herds." The branding iron honors marketing leaders, and the "cornerstone" recognizes those with superior dedication to product integrity. At the 2008 gathering, conference attendees vote live for the recipients of the brand's first People's Choice trophy (right), "The Cowboy." The inaugural winners are Piggly Wiggly Carolina for its Valentine's beef campaign, along with Southern Foods and Sysco Spokane for their sales team contests.

Efforts by the brand's media relations team help licensees gain front-page play in local and regional publications. Since 1996, according to *Google News*, more than three thousand articles have contained the exact phrase "Certified Angus Beef" and without the quotations—which ensures the words are used consecutively in an article—there have been 24,500 mentions of Certified Angus Beef.

In October 2001, *City Visitor* magazine, which reached four hundred thousand in Ohio, wrote an extensive feature on steakhouses. The story included John Q's Steakhouse and an interview with Stoffer.

In 2003, there was a big push on promoting Certified Angus Beef recipes. One was picked up in the *The Tennessean*, the Nashville newspaper with a circulation of more than 181,000. That same year, a Ginger Roast Salad recipe and other recipes were picked up by the *National Enquirer*, which has a circulation of nearly two million.

Such top-tier hits continued throughout the 2000s, as brand awareness grew. In June 2007, Sharper Image featured Donald Trump on its cover. The story announced the exclusive introduction of Trump Steaks, Certified Angus Beef Prime.

In the story, Trump said: "For years, people have told me the steaks in my restaurants are incredible. So it makes sense that we expand our business to offer Trump Steaks to consumers everywhere. The Sharper Image is one of my favorite stores with all kinds of fantastic products and is the perfect venue for announcing to the world that Trump Steaks are ready to be enjoyed at home with friends and family." The story was mentioned by TV hosts Jay Leno and David Letterman.

The publicity added to great brand awareness. According to extensive consumer research, 87 percent of consumers recognize the brand name, with the logo having 91-percent recognition. That was 3.5 times greater than the next closest brand, says the 2007 branded beef study conducted by Shiller, Polo and Benson for Certified Angus Beef.

In 2008, the big publicity winner in the trade and consumer media was the brand's thirtieth anniversary celebration at the James Beard House, considered the Mecca of U.S. culinary arts. *The Virginian Pilot*, *The Cincinnati Enquirer*, and *The Akron Beacon Journal* wrote about the event. For licensee Chef Michelle Brown of Jag's Steak & Seafood near Cincinnati, Ohio, it was an honor to cook there.

"Cooking at the James Beard House was an amazing experience," said Brown in the news release. "It was an honor to work with the team of chefs and to experience the camaraderie, the knowledge, and the teamwork it took to put on a world-class event. It was a privilege to be selected, and the evening is one I'll always remember."

Brown's story was featured in the Cincinnati newspaper with an estimated readership of two hundred thousand, plus an online audience.

The brand also hosted the media for a dinner at the Bull and Bear steakhouse. Board members—cattle producers—were invited to showcase the strong ranching heritage of the brand. President John Stika and Board chairman Bill Davis were interviewed live on *Fox Business News*.

2000

Suppliership to 2002 Olympic Winter Games in Salt Lake City announced; featured products are frankfurters, barbecue beef, pot roast, and deli meats; chili added later

First international target market: Vancouver; promotions include sponsorship of *Vancouver Sun* Run

Program reorganized into Limited Liability Corporation

Spring "Here's the Premium" study reports $10–15 per head advantage for Angus calves over other breeds

First large packer offers Certified Angus Beef brand Prime: National Beef

Since 1989, educational programs have taken customers to the farm to understand the brand and the dedication of Angus ranchers. At the 1990 Boot Camp (right), Angus rancher Jim O'Neill leads seventy people on a tour of his family's Iowa ranch. Sessions for the same seminar, now called Roundup, include presentations from meat scientists like Dr. Daryl Tatum, from Colorado State University (above).

Education Initiatives

Every year, Certified Angus Beef's annual conference is the brand's biggest event. Officials name licensees of the year in marketing, brand integrity, and sales categories and give out individual awards. A news release is written and sent to industry and local publications. For example, Advance Food Company won Top Sales Volume Processor, Value-added Products Processor Marketer of the Year, and Top Processor Sales Volume Increase in 2008 and the following year. The 2008 news release was picked up by the company's local paper, *Enid News & Eagle*.

Advance Food Company marketing coordinator Pam Godlove says the annual conference, which has featured big-name speakers like legendary NFL coach Don Shula, "helps everybody put our best foot forward."

Since the mid-1980s, when distributor representatives met in those small hotel rooms, these meetings have been the lifeblood for Certified Angus Beef's many licensees. It's the only venue, in all the industries, where competitors meet, collaborate, and share ideas about how to make Certified Angus Beef work. In a way, it's like Walmart and Target getting together to talk about what each can learn from the other on how to market a line of products.

Instead of a competitive atmosphere, it's a friendly one. And it's always been that way, says Oxford's Charlie Robinson, who has been to nearly every annual conference.

"There weren't a lot of people in the early days, but we had a few licensees, National Beef and myself," Robinson remembers. "At these meetings, you really felt like you were doing something for somebody."

Paul Mackay says he has made numerous connections that have helped him improve his El Gaucho steakhouse in Seattle. As his steakhouse grew in popularity and the

In recent years, Certified Angus Beef has extended its "on farm" education to include more chefs, retail meat managers, and media. American Angus Association member Phil Trowbridge (left) hosts eleven reporters at his Ghent, New York, ranch. They discuss cattle feeding, environmental stewardship, marketing claims, and marbling importance.

Certified Angus Beef brand continued to grow in his market, Mackay wanted to upgrade to Prime to separate himself from the marketplace. At the time, Certified Angus Beef had not officially extended its brand to Prime, but he had heard about a packer who was separating Certified Angus Beef carcasses that qualified for Prime and selling them at a premium. At one conference, he met with John Tarpoff of Tarpoff Packing, who helped champion the Prime extension. "We had a cigar and I said, 'John you're the guy I heard about that has the packing plant and this incredible eye for beef. Would you sell me the Prime that also qualifies for Certified Angus Beef?' "

Some time later, the Prime line was launched and Mackay had his wish, but the conference gave him the face-to-face time he needed to talk about this important subject. Mackay also picks the brains of steakhouse managers around the country. It's a great venue for him to learn about the full scope of the business, he says.

On a smaller scale, the brand executes Building Blocks seminars for licensees at least three times a year. The program brings producers and distributors together to educate them about the science behind the brand and the integrity by which it's kept. Certified Angus Beef takes a grassroots approach to education. Over the years, these efforts have trained thousands of waitstaff, retail meat managers, chefs, distributors, and consumers.

Since 1998, Certified Angus Beef has focused on educating the licensees' employees. At Boot Camp in 1989, the drill sergeants were Colvin, Stowell, and VanStavern, and featured a presentation from Houston restaurateur Hendee. Boot Camp, later called Roundup, was such a success that it was held again in 1990 at Nebraska Boxed Beef in Omaha and the O'Neill Angus Farm in Logan, Iowa. It brought together distributor sales representatives from all over the country. They learned about beef fabrication,

2000

New consumer campaign: "What vegetarians eat when they cheat" launches in St. Louis and Boston

Beef Product 2000 competition brings national acclaim to four Certified Angus Beef products; RMH Foods top sirloin in bourbon sauce and meatloaf; Bubba Burger patties; W&G Marketing barbeque beef

Certification begins in Canada at IBP-Lakeside (later XL Beef)

Annual sales surpass 500 million pounds

More than 10 million Angus-influenced cattle evaluated during year; 48 percent of cattle in licensed plants are black-hided

First restaurant offers prime steaks: El Gaucho

"Certified Angus Beef" brand name receives U.S. trademark registration

USDA grading, and life on the farm. For many of the sales reps, it was the first time they had been to a cattle operation. They went from zero knowledge to a good understanding after a crash course in cattle production and Angus genetics 101 taught by Colvin and company.

In 2003, that company launched a CD-based training tool, ORBIT (Official Resource for Branded Beef Industry Training), to enhance the ability of foodservice sales reps to present to potential customers. The waitstaff training video from 1997 has evolved into an interactive kit, filled with an instructional video, lapel pins, reference cards with reasons to order a Certified Angus Beef steak, and an informative poster that compares the brand to USDA Choice.

Training for retail meat managers has been another mainstay of educational programs. In 1995, "Creating Consumer Confidence" actually created retailer confidence with cut selection, cooking guidelines, and recipes. Expanded to a CD-based program ten years later, these tools help retail staff recommend cuts to customers.

Recently, there has been a concerted effort to bring more distributors, retailers, and chefs back to the ranch much like the original Boot Camp days. Feedlot manager David Trowbridge has helped with this initiative. In 2007, some Las Vegas chefs visited his Gregory Feedlot near Tabor, Iowa, to understand the process of producing what they put on a plate.

"When they come into the office, they have to 'buy' a calf as a demonstration," Trowbridge says. "We go through the different genetics and management programs these cattle were on, and then we feed the cattle out. Some get sick, some grade Choice, and some make Certified Angus Beef."

Trowbridge then works with the chefs to look at the economic risks and the number of factors that can hinder profits. He says the chefs always find the process fascinating: "They look at all you do and say, 'You take all that risk and hope to make only $33 a head.' "

The brand forms a strong partnership with Amy Tobin, a rising food personality based in Cincinnati. She makes guest and media appearances at consumer events and often recommends the brand to followers on her website and syndicated radio show.

Above: A promotional partnership with Weber-Stephens Products Company, maker of world-renowned Weber grills, is a natural fit. At consumer events, chef and cookbook author Jamie Purviance shares tips for grilling a perfect steak, starting with Certified Angus Beef and a Weber grill. The grills are also given away during summer promotions at licensed retail stores.

Left: Joint promotions with complementary premium brands create win-win results for the companies, licensed businesses, and ultimately consumers. In summer 2008, Tripps restaurants introduce steak Caesar salad and ribeye entrées through a joint promotion with Lea & Perrins Worcestershire sauce.

Even with all the marketing efforts and focus on delivering a great eating experience for consumers, it's ultimately about driving demand for registered Angus cattle. The breed is known for both their maternal and carcass traits including an inherent ability to marble.

Photo credit: Thomas Angus Ranch

11 Ranch Partners

When the supply development division was formed, the goal was to increase the number of available animals through the use of better genetics and feeding programs. Staff hoped there was enough buzz about Certified Angus Beef that feeders would send their cattle to licensed packers. But it was a tough sell in the late 1980s.

"There weren't really a lot of programs available that connected the producer to the packer," says John Stowell, the first supply development director.

To begin program development, Stowell started feeding and carcass contests, working with a Kansas feedlot and its packer-buyers. Typically, each producer consigned five to twenty black-hided steers, which were fed together but recorded separately at harvest. At that time, Stowell collected data and marked the ones accepted for the brand. In 1990, the producer who had the most qualified cattle won $5,000, but the real trophy was non-stop "bragging rights" over his fellow producers, Stowell says. Winning also added long-term value to a herd proven to produce the highest quality.

In 1991, Larry Dorsey replaced Stowell and took supply development to the next level. Dorsey and staff coordinated carcass data collection at packing facilities. They would then trace the data back to the sires in an effort to improve supply lines. Dorsey also worked with cattle producers who were collecting their own carcass information and told their stories through the *Angus Journal*. Certified Angus Beef vice president Larry Corah credits Dorsey with shaping the future of the company's supply development program and for looking at the division as a long-term means of making marbling part of the genetic focus. Former associate director Mary Ferguson agrees.

During the first Value Discovery Project, supply development director John Stowell (left) talks with Angus cattle marketer Cal Siegfried at Decatur County Feed Yard in Kansas. This program educates producers about their herd genetics by using carcass data. The Angus Feeder Cattle Directory *(above) helps market those cattle to feedlots.*

"Larry Dorsey is the person most responsible for growing the carcass EPD [expected progeny differences] database and developing it into a tool sought by Angus breeders today," Ferguson says. "He had a great rapport with Angus breeders and commercial producers and, as a result, was extremely successful in coordinating sire progeny tests for carcass merit."

Corah explains, "The data Dorsey collected would be turned over to the American Angus Association and become part of the database that breeders use today."

Dorsey also picked up where Stowell left off in terms of helping cattlemen evaluate their herds, with the Value Discovery Project in 1992. Decatur County Feed Yard at Oberlin, Kansas, continued its partnership with Certified Angus Beef to hold the carcass contest. Many producers entered just to gather information about their cattle. Garden City, Kansas, cattleman Sam Hands told the *Angus Journal* the project was ideal for learning how cattle fit into the Certified Angus Beef program.

"I want to find out the genetic capabilities of the cattle in relation to the program specifications," said Hands, whose Triangle H Grain and Cattle Company would later become the first licensed Certified Angus Beef feedlot. "Our goal is to harvest cattle at a younger age and still produce well-marbled carcasses, so we need to identify sires that can help achieve this."

In 1998, the brand took its outreach to producers to a new level. Ferguson had been ranching and served only as a part-time consultant since 1992, but she came back for a six-month stint as acting director of supply development. She raised the effort into a three-pronged approach that would invest 10 percent of the company's budget into helping producers understand how and why to hit the Certified Angus Beef brand target.

Her proposal to focus on progeny testing for carcass merit, industry information, and feeder-packer relations met unanimous board approval.

The feeder-packer relations division created a Feedlot Licensing Program, since virtually all cattle move from ranch to feedlot to packer. The move generated much more commercial carcass data and broadened the availability and interest in such information that quantified marbling.

"We worked with a lot of scientists around the country to put marbling data together that we could communicate to the beef industry," Corah says. "For literally twenty years, Mick had producers come up to him and say, 'Mick, I like your branded beef program, but where's the premium for producers?' "

Above: Beginning in 1991, Larry Dorsey continues efforts to gather carcass data and improve supply lines. He also encourages producers to evaluate and choose sires based on beef quality factors. They begin to make marbling part of their genetic focus.

Left: In many respects, this "Building Tomorrow Together" brochure invokes the cooperative approach that has made the brand so successful. The company has been a liaison between producers, food companies, and ultimately consumers.

No longer would Certified Angus Beef staff be at a loss when that question was posed. Through the annual "Here's the Premium" study, the company showed proof of an increasing premium for Angus over non-Angus calves. The 1999 premium of $15 per head grew to $34 per head by late 2008. In 2003, packers paid more than $35 million in premiums for Certified Angus Beef and by 2008, the cumulative amount had surpassed $250 million. Furthermore, in 2008, USDA identified about 14.2 million Angus-influenced cattle at harvest, or about 60 percent of all finished cattle harvested in licensed plants that year.

Above: Sam Hands (right), Triangle H Grain & Cattle Company, signs the first feedlot agreement in 1998 with new company vice president for supply development, Larry Corah.

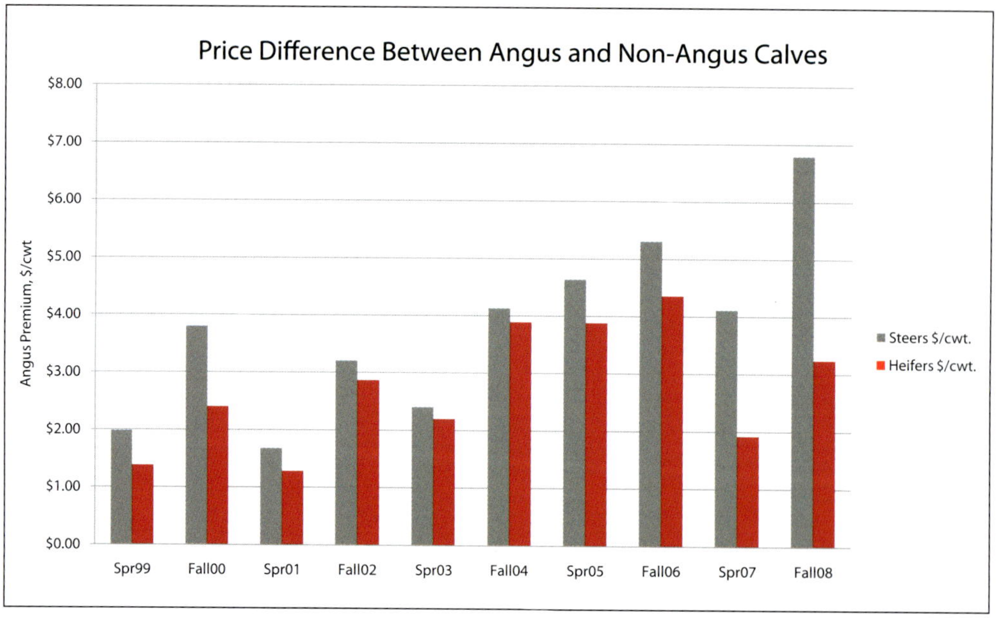

Right: For years, producers seeing the success of Certified Angus Beef asked, "Where's my premium?" The "Here's the Premium" study answers their questions. Research with auction markets across the country documents the Angus advantage over non-Angus cattle, and the premium doubles in ten years.

The Feedlot Licensing Program helps cattle feeders gather information and better understand Angus genetics. Company representatives work with their licensed packers to collect carcass data on cattle from licensed feedlots. Then they help the cattle feeders present the information to their cow-calf customers so that all can cooperate in producing better Angus cattle, more profitably.

Following the example set with the restaurants and retailers, Certified Angus Beef offers feedlots the brand's marketing power. Feedlot licensees gain use of the trademarked logo with "Partner" in block letters at the base of the traditional Certified Angus Beef mark. They use it on their websites, billboards, jackets, sales sheets, and other marketing materials. If their cattle go through a licensed feedyard, cow-calf producers can market the association with Certified Angus Beef as well.

For Gregory Feedlots, a licensed feedlot since 1998, Certified Angus Beef has been crucial in helping teach customers about genetics.

"We encourage our producers to use the right genetics," says David Trowbridge, manager of the Tabor, Iowa, feedlot. "We've changed a lot of producers' breeding programs."

Thanks in part to licensed feedlots, more Angus cow-calf producers had more success hitting the brand's target and getting paid for doing so. That filtered down to ranchers who sell their calves at auction, too.

"The brand has done more than the original founders probably thought it would," says Hands.

The feedlot program quickly became a point of differentiation. If a business was a partner in the Feedlot Licensing Program, that signified to cow-calf operators that the company looked at cattle as more than a commodity; they were shooting for the top premiums. In less than two years of the program's existence, Certified Angus Beef had licensed fifty feedlots and 55,357 head were enrolled. The brand never intended to license a significant share of the feeding industry, but about one hundred licensees had built a detailed feeding and harvest database of 1.5 million records as of 2009.

"Black Ink Basics" seminars help Angus producers focus on profit and quality. More than 150 cattlemen and women attend this event in Texas, learning how to better manage and market Angus calves.

Industry Information

Steve Suther, director of industry information, and his team keep Angus producers informed about how to improve their herds and how Certified Angus Beef is helping them. Suther's division writes a monthly column, "Black Ink," and distributes it to all agricultural journals.

The work is frequently picked up by every major industry publication because of the staff's long-time relationships with the agricultural media and because they provide valued articles and reports for free. While the pieces have always been newsworthy, they often promote the agenda of Certified Angus Beef in a way that makes producers want to be affiliated.

Suther, a career agricultural journalist and former *Farm Journal* staff editor, also helps manage the family's commercial Angus herds in Kansas. "That background and his writing ability have helped in our outreach to producers," says fellow staffer Christy Johnson.

In a 2002 column that ran in *BEEF*, Suther wrote about cattlemen who are tired of seeing so many black cows. To them, he had this advice: "Breed them to registered Angus bulls with the industry's most accurate expected progeny differences (EPDs) to produce calves that can hit the premium Certified Angus Beef target. Then, watch to see if your cows have what it takes to carry those bloodlines to fruition."

In addition to "Black Ink," Suther and his team produce radio features and write for the *Angus Journal* on a monthly basis and distribute cattle-oriented stories throughout the industry each week. The majority of their work can be found at *CABpartners.com/news*.

2001

Little Apple Brewery wins National Beef Backer Award

Thirty-Aught (later 30.06) Program recognizes feedlots for high-quality cattle; Heartland Feeders II wins first "bronze" award for 500 cumulative cattle with 30 percent or more qualifying for brand

Urner Barry publishes first Certified Angus Beef Market Report in daily *Yellow Sheet*

Consumer research leads to new premium logo, which gains U.S. registration in 2003

Brand Builder sessions in Texas, North Carolina, Georgia, and Florida foster producer ownership and understanding of program

Gregory Feedlots owner Jim Gregory (left) and manager David Trowbridge say being a licensed partner attracts more high-end customers with better Angus genetics.

Company president John Stika, a former director of supply development, talks about the days when feedlot licensing was just beginning. "Carcass data wasn't flowing and still doesn't flow freely," he says. "But what we're trying to do is gather data on as many Angus cattle as we can, to try to come up with a tool to help us figure out how to manage these cattle more effectively for Certified Angus Beef."

Typically, the animals were sold on "cash" terms, which meant the feeder forfeited rights to any information passed, beyond the amount on the check. The packer owned

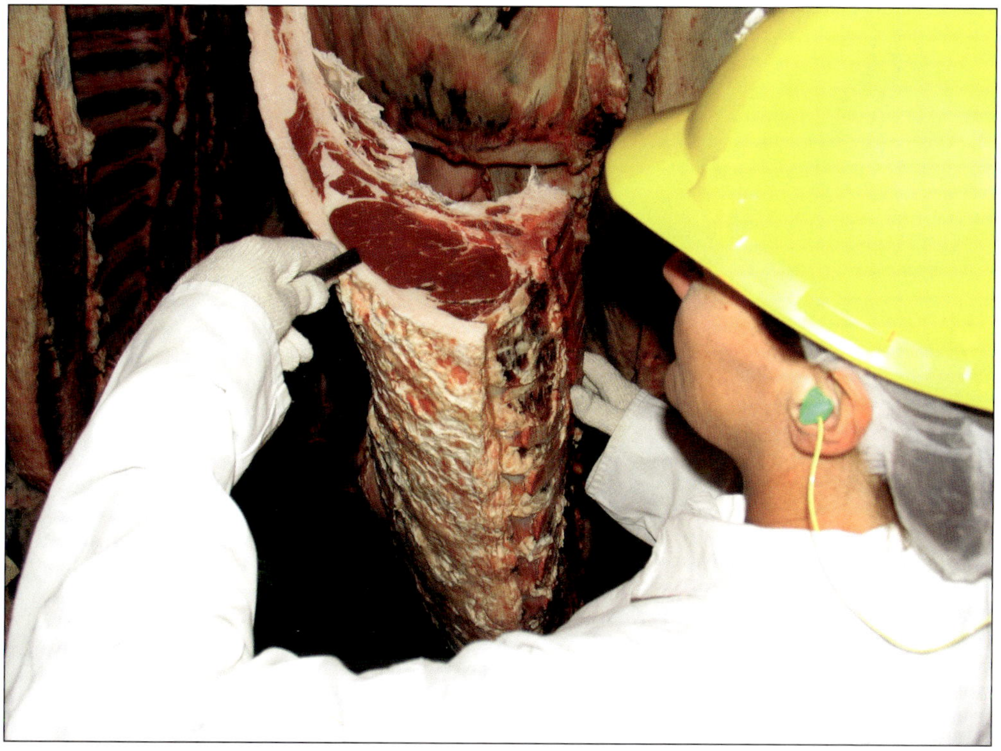

Thousands of cattle from more than a dozen states come to the Tri-Country Steer Carcass Futurity in southwest Iowa so their owners can learn from the feedlot and carcass data. Records help producers understand how well their cattle meet demand for quality and consistency.

all the information about the cattle. Although there were some successful data collection businesses out there, Certified Angus Beef joined the effort. "We were trying to allow the feeder access to information that the packer was not obligated to provide," Stika says. "We didn't realize how much we were trying to change, but to get data at the volume we did from as many different packers was a fairly big success for the program."

From a packer's perspective, Certified Angus Beef has led this initiative.

"Supplying carcass characteristic information back to Angus breeders and the Certified Angus Beef team has been an ongoing effort over the years," says Bill Buckner, Cargill CEO. The packer gains by improved carcass performance in fabrication "while continuing to provide a quality eating experience for consumers."

According to feedlot partners, the program has been as important to them. Trowbridge says being licensed gave his clients an opportunity to find a niche market and receive a higher premium. In addition, the partnership has given Gregory Feedlots a lot of advertising, which has attracted "more high-end customers."

Because Gregory's pens are filled with cattle from numerous farms, there's variation in how many qualify for the brand. There will be times when one pen of cattle has a higher acceptance rate than the other, but both pens were on the same feed and treated exactly the same. When that happens, Trowbridge tells the lower-qualifying producer why: "It's genetics. We've changed a lot of producers' programs by what they'd seen from other producers here. Some of them were pretty slow about doing it, but if they want to get those premiums, they will change."

2001

Usinger's Certified Angus Beef frankfurter named "top dog" in taste test co-sponsored by Baseball Hall of Fame

Nearly 1,300 stores display "Go for the Gold" point-of-sale materials leading up to Olympic Games

Products available in more than 4,300 restaurants and 4,000 grocery stores

Website for producers launches (*CABpartners.com*)

First licensed restaurant in Middle East: Fuddrucker's

Advance Food Company creates chili for Salt Lake Games; other new items from the company include fully cooked burger and country-fried steak

2002

CattleFax determines demand created by brand added more than $450 to value of every Angus bull sold by Angus seedstock producers in 2000

Suppliership to 2002 Olympic Winter Games leads brand to world stage; more than 1.5 million served, and more than 100 frankfurters per minute

Tyson Fresh Meats begins identifying teres major, one of brand's first value-cuts

Retail promotions include Kikkoman soy sauce, Birds Eye vegetables, Kendall Jackson wines, and Broil King grills

Bull and Bear in Waldorf=Astoria licenses for Certified Angus Beef brand Prime

At Angus University classes in 1999 and 2000 in Montana and Missouri, some 270 ranchers participate in the brand's first, largest, and most in-depth educational program for producers. They learn how to analyze muscle and marbling on live cattle using ultrasound; meat scientists discuss differences in beef cuts.

And when Trowbridge has a customer giving anecdotal arguments about changing genetics—whether they are narrowly focused on a certain bull or feel like they have to buy from their brother's herd—all he has to do is show the data. Because the information has the Certified Angus Beef logo attached to it, it has credibility, so they know the feedlot isn't using junk stats.

When the company talked to Sam Hands about licensing his feedlot in 1998, he had already been practicing carcass data sharing with his clients and recommending Angus genetics. But he jumped on the chance to work with the brand.

"It gives you more marketing power. There's no question about it: When you have any association with Certified Angus Beef, you're that much higher on the ladder," he says. "I am grateful to be a part of the program. Whether you're a restaurant or a rancher, the affiliation says a lot."

For Gary Darnall, whose Darnall Feedlot consistently ranks among the top performing programs, the affiliation with Certified Angus Beef enhances the opportunity to feed higher-quality cattle. Darnall Feedlot won the National Angus Carcass Challenge heifer division in 2004 with 50 percent Prime and Certified Angus Beef. Darnall's steers came in third that year, but the yard fed fifty-five pens over the five years. Gary Darnall told the *Angus Journal* that all the animals from his ranch came from similar genetics. He conducts ultrasound sorting eighty days prior to harvest at his twenty-thousand-head feedlot to aid in hitting the target.

Kansas State University's Ag Research Center, in Hays, wins the 2002 Best of the Breed contest with Angus genetics drawn from local ranches. The eighty ultrasound-selected steers achieve 91 percent acceptance for the Certified Angus Beef brand.

"We have a desire to be a leader in our industry, but the reason we've received awards was the advantage of feeding good-genetic cattle from our own herd and other ranchers' herds," Darnall says. "We have been fortunate to have access to those cattle and customers that have that quality cattle; Certified Angus Beef has helped us get them because people want to be affiliated with the program."

Strategic Brand Partners

Just as the company's marketing division partners with high-quality companies, so does the supply development division.

"We've partnered with a lot of people, and because of our not-for-profit financial structure, we're a very easy group to partner with," Corah says. "Through these partnerships, we try to create the pull-through effect on the producer side just like we do on the product side."

One of the first companies to step up to help with "Quality Assurance" training was Fort Dodge Animal Health. Frequent partners include Pfizer, Purina, *Drovers* magazine, and *Feed-Lot* magazine. Starting in 2006, Certified Angus Beef, Pfizer and *Feed-Lot* magazine have hosted annual Feeding Quality Forums from Texas to Nebraska in the fall. Some one hundred and fifty feedlot managers, nutritionists, educators, and industry professionals attend, learning about trends, economics, consumer focus on marbling, growth technologies, and much more. The information-packed sessions tell of the many ways science is having an impact on the industry.

2002

H-E-B Mexico licenses 30 stores, and by 2008 becomes largest volume retailer for chains with fewer than 50 stores

Regional sales team establishes customer service links throughout country

In June, July, and August, monthly sales exceed 50 million pounds

Research shows 70 percent of consumers recognize brand; of those recognizing it, 87 percent buy it, and nearly 65 percent do so because of reliable quality, taste, and appearance

Carcass Data Collection Service discontinued with industry adoption of ultrasound; it generated more than 150,000 data entries and helped producers evaluate 5,000 bulls for carcass merit

Half of fed cattle in country black-hided

First premiums for cattle from licensed feedlots: JBS Swift in Dumas, Texas

Feedlots

Running a feedlot is a risky business. A cattle hotel before harvest, feedlots make their money off of room and board as well as their feeding program. With the price of corn and other feeds constantly changing, it's not an easy business in which to make money.

Furthermore, they're judged on how well their cattle grade and in many feedlots, they do not even know their cattle's genetics.

When cow-calf producers take their calves to public auction, anonymous bidders trade bids. Most of the time, "order buyers" purchase the animals. The commissioned buyer then contacts his clients—typically feedlots, investors, or stocker operators—but often does not know anything about the cattle he has just purchased other than the obvious.

"Sometimes a stocker operator or the feedlot will have their own procurement people at an auction to buy direct," Suther says. "In addition to all these auction sales, there's a fair amount of ranch-direct business where order buyers or agents for stocker operators go around the country and look at groups of cattle for sale."

Certified Angus Beef works with licensed feedlots to help them meet challenges and remain profitable enterprises. As market trends and consumer demands shift, the brand holds firm to its mission to improve the nation's Angus herds.

Suther and his industry information staff have long encouraged producers and feedlots to share information about their cattle. He says most cow-calf producers do not just pay feedlots to manage their cattle.

"In our ideal world, that would be much different as it would ensure most cow-calf producers took a personal interest in post-weaning performance and carcass value," Suther says. "In reality, most producers sell their calves at or shortly after weaning, and most of those calves become 'commodities' in the sense that they are intermingled with calves from other farms and owned by feedlots or stocker operators or investors and fed to the best ability of the feedlot to make money. We often encourage retaining ownership of one's calves in a feedlot, but there is risk in long-term ownership. It can be lessened by a focused effort to improve cattle and get paid for added value, such as excellent quality sold on a grid."

Above left: Diane and Mason Fleenor (fourth and fifth from left), licensed feeders in Iowa, win the 2003 National Angus Carcass Challenge. Their forty-four GG Genetics steers graded 11 percent Prime and 89 percent Choice, with 59 percent Certified Angus Beef acceptance.

Above right: Nebraska's Darnall Feedlot becomes licensed in 1999 and quickly earns awards for volume of qualifying cattle. Darnall uses ultrasound and all available data to improve Angus genetics and clients' premiums. With his son, Lane, Gary Darnall (pictured) has become a valued partner.

Through "Consumer First Beef Partners," Pfizer, Purina, and *Drovers* joined with Certified Angus Beef in 2008 to build bridges between consumers and producers. A round-table "Think Tank" discussion that fall featured such industry leaders as Daryl Tatum, meat scientist at Colorado State University; Jim McAdams and Mike John, former presidents of the NCBA; Tyson's Mostek; Sysco's Malcolm; and food industry consultant Dan Murphy. The group identified ways to better meet consumer demand for high-quality beef.

In addition to events like the Feeding Quality Forums, "think tanks," and sponsored white papers, Certified Angus Beef contributes to research when it can. Corah says the brand follows the lead of its parent company.

"We're looking at how to incorporate DNA into making genetic information more predictable—that's under the leadership of the American Angus Association. We're very interested in not only its capabilities of enhancing genetic predictability in the near term, but we want to visualize production systems of the future," Corah says. "We could be sorting cattle into outcome groups based on their DNA profile, and one of those would be a quality-based outcome group. That's an exciting concept and it has widespread applications, including on the genetic side. We will do whatever we can to assist in that development."

AngusSource

In October 2005, USDA approved the Association's AngusSource® program. The first of its kind, it is a USDA Process Verified Program that documents source, group age, and minimum 50 percent Angus-sired genetics.

When packers see the AngusSource tag, they know those animals are more likely to qualify for Certified Angus Beef or Prime. Thus, they may receive more dollars for the

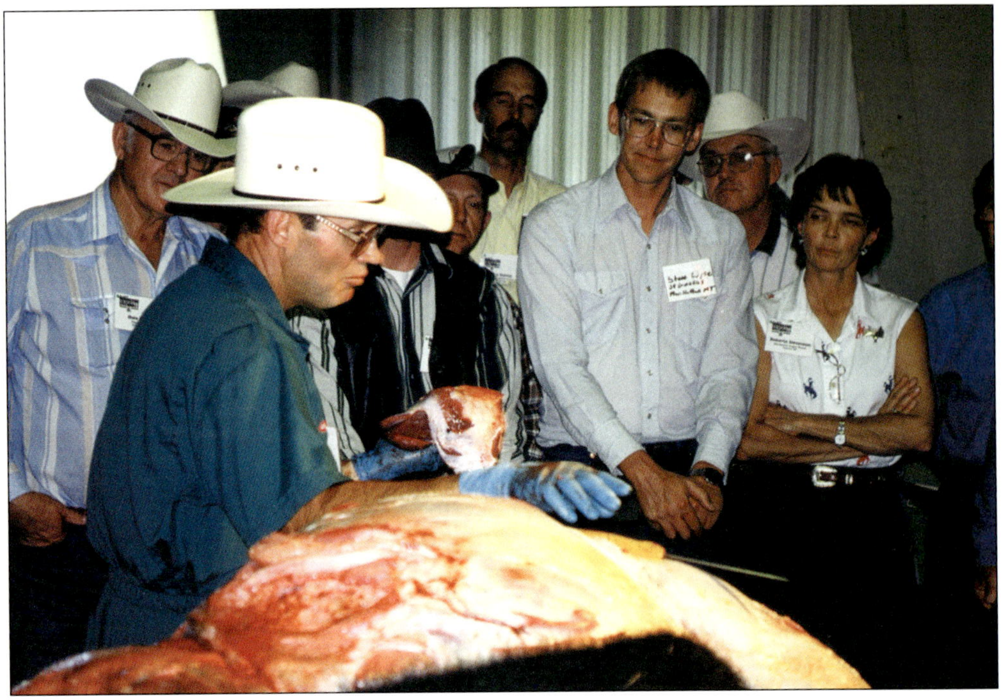

Above: Tom Williams licenses Chappell Feeders in 2004 and quickly stands out as one of the best. Within four years, he feeds more than 2,000 customer cattle to hit the "golden" benchmark level of at least 30 percent brand acceptance with few discounts.

Above right: Quality Assurance Seminars with allied beef industry partners help train feedlot personnel. Dr. Jerry Woodruff (forefront), veterinarian for Fort Dodge Animal Health, shows a class how ranch and feedlot practices affect beef quality, and how to preserve its full potential.

carcass and are willing to pay more. At the end of 2009, more than 415,000 cattle had been enrolled in the program.

In reference to Tyson sorting AngusSource cattle, John Butler, chief executive officer of the Beef Marketing Group (BMG), told *CattleToday.com* that the program rewards "cattle with good genetics, cattle with good feeding performance, and cattle with excellent carcass performance as well. That's a huge advantage for us."

Bryce Schumann, the Association's CEO, says AngusSource allows the industry to segment black-hided cattle within the marketplace.

"In a lot of ways, we're putting out our seal, our verification, that these cattle are in fact sired by a registered Angus bull. Besides documenting those genetics, we're also documenting source and age," Schumann says. "Age is a critical factor, so being able to document those cattle that are twenty months or less of age, there are some premiums available in the marketplace for that. AngusSource allows people to document a lot of third-party, verified information that we can get out into the marketing channel to assist people using Angus genetics in getting more dollars for their genetic investment."

Reaching Out to Producers

Christy Johnson, part of the supply development team since 1997, says producer education has been a key strategy since the 1998 board decision to invest in that area.

At the educational seminars, producers learned and interacted with long-time licensees. According to the *Angus Journal*, they also learned about necropsy and injection-site analysis. Fort Dodge Animal Health veterinarian Jerry Woodruff examined a carcass to review a poorly managed animal, pointing out the beef-quality implications. Producers learned everything

from the value loss associated with the processing errors, to proper needle size, use, and application. Johnson says these messages indirectly affect Certified Angus Beef supply.

To further educate producers about the brand, in 2001, Certified Angus Beef began holding Brand Builders seminars, which, like Building Blocks, educated producers about the brand and how targeting consumers supports the price of bulls.

"Having Angus producers in our office is a tremendous opportunity for everyone," said McCully, currently assistant vice president for supply development, at the first seminar. "The brand is theirs, and when Certified Angus Beef is successful, they are successful. It only takes looking at Angus cattle prices to realize that."

In recent years, Certified Angus Beef has also sent its beef specialists to ranches and feedlots to help everyone along the production chain better understand health and management that make the most of Angus genetics. But the top producer outreach program in the company's history has been the 2007 *Best Practices Manual*, Johnson says. It's five chapters long and covers general management, genetic selection, health, nutrition, and marketing.

The manual, carcass contests, and meetings have become increasingly valuable to producers over the years, supplementing university Extension programs that cannot focus

In 2006, AngusSource is added as a live animal specification for qualifying cattle. It is a USDA Process Verified Program that documents source, age, and a minimum 50 percent Angus genetics. Having enrolled his calves, Kansas rancher Luke Hoeme (left) and Nebraska licensed partner Bill Robinson, Highland Feeders, discuss sharing ownership and performance data.

At a Brand Builders meeting, Larry Corah (head of table), vice president for supply development, speaks with producers and feedlot managers about improving supply lines. This is one of many outreach programs Corah has established to open communications with producers.

so much attention on Angus cattle alone. McCully says producers are looking for recommendations on nutrition, marketing, and genetics.

"They don't have a lot of options; it's hard to take recommendations from a feed salesman alone, when he's just trying to sell you his feed," McCully says. "Producers look at us as a third party. Our role is not to push something up the system; it's to make sure the market signal for quality is finding its way to the ranch, and make sure the decision-making tools are in place."

One of the decisions producers have had to face in the past ten years has been whether it is possible to aim for both production efficiency and beef quality at the same time. Certainly, tools such as the *Best Practices Manual* help sort those ideas out, but first, the groundwork was laid with a series of scientific white papers.

Corah led industry "think tanks" from his earliest days with the brand, exploring ways that consumer demand for quality could better reach producers and affect their selection and management. A score or more of those white papers continue to draw producer interest on the brand's *CABpartners.com* Web page. Johnson says the most popular of those has been the jointly authored Corah-McCully title, "Declining Quality Grades: A Review of Factors Reducing Marbling Deposition in Beef Cattle," and their 2009 sequel, "Quality Grades: What's Driving the Recent Upswing?"

In addition to the constant commercial concerns, the "Natural" and "Prime" issues arise—indeed they begin—on the ranch.

When Certified Angus Beef debated these product lines, the supply development staff frequently assured producers that these were not in competition with the traditional branded products.

"The Prime and Natural extensions let us develop products that are in demand at the retail and foodservice level," Corah says.

To increase acceptance rates for cattle qualifying for Certified Angus Beef Prime, the staff communicates a strong strategic message of managing nutrition and improving genetics. There's also the economics communiqué: Cattle good enough to qualify as Certified Angus Beef Prime bring their owners more money per pound than 99 percent of other cattle.

"We have some producers doing an amazing job in the percentage of Certified Angus Beef Prime they create," Corah says.

Since 2004, when the Natural line became available, Certified Angus Beef has worked with suppliers and packing partners to help change production practices for producers interested in supplying that product extension. The staff helps traditional producers adapt to the practices without losing efficiency.

"I'll have to admit, it's been amazing to me how successfully natural product can be produced in the beef industry," Corah says.

Mr. Chairman

The chairmen and members of the Certified Angus Beef Board were very influential to the brand's development. They brought their expertise and viewpoints to the table, making the brand and the breed stronger in the long run.

CHAIRMEN OF ANGUS BEEF CERTIFICATION COMMITTEE

1975–76: Gilman Stewart, Stewart Select Angus, Greensburg, Indiana

1977: Mick Colvin, American Angus Association representative, West Salem, Ohio

CHAIRMAN OF CERTIFIED ANGUS BEEF COMMITTEE

1978–84: Fred H. Johnson, Summitcrest Farms, Summitville, Ohio

CHAIRMEN OF CERTIFIED ANGUS BEEF BOARD OF DIRECTORS

1985–87: Harvey Lemmon, Lemmon Angus, Woodbury, Georgia

1988: Wayne Wyble, WE 7 Angus Farm, Opelousas, Louisiana

1989: Dick Janssen, Green Garden Angus, Ellsworth, Kansas

1990–91: T.D. Steele, Mill Brae Ranch, Maple Hill, Kansas

1992–95: Dr. R.J. Hillier, Hillier Angus Ranch, Stillwater, Oklahoma

1996: Bill Rishel, Rishel Angus, North Platte, Nebraska

1997–98: Henry Bergfeld, Summitcrest Farms/Pine Hill Farms, Summitville, Ohio

1999–2001: Steve Brooks, Brooks Chalky Butte Ranch, Bowman, North Dakota

2002–06: Robert Norton, Jr., Biozyme, St. Joseph, Missouri

2007–08: Bill Davis, Rollin' Rock Angus, Sidney, Montana

2009: Rob Thomas, Thomas Angus Ranch, Baker City, Oregon

Twenty-two years after he took the job of executive director, Colvin stands in front of company headquarters in Wooster, Ohio. Opened in 1994, the office symbolizes all he has accomplished, from selling the first pound to nearly 500 million pounds in his retirement year of 1999.

Photo credit: Sheryl Smith-Rodgers

12 Replacing a Legend

As evident from the great strides bull prices made and incredible brand growth during his tenure, Mick Colvin was the right man for the job of Certified Angus Beef executive director. In the early days, he endured personal insults from people who didn't like the fact that he was changing the industry. He got caught up in a political whirlwind that first year, when USDA pulled the plug on the brand. And he had to bear two years of a grueling court case that could have put the company at risk. But Colvin endured.

Even when distributors and packers laughed at him, saying his Certified Angus Beef program was a sham, he got up and tried again. The man faced every challenge with an unrelenting passion and a heart of gold. Most people would have quit and moved on to something more promising. But not Colvin.

"I love challenges," he says.

For nearly twenty-two years, he managed Certified Angus Beef and developed a working culture that would be envied by every outside company leader who walked through those Wooster, Ohio, doors. When he retired from the post, he didn't walk away a wealthy man, but he left making a huge difference for the Angus breed. But you will never hear Colvin admit to such an accomplishment. If anything, the magnanimous Colvin would give credit to everyone else on the team.

It's that kind of appreciation for others that made Colvin a special person for former employee Chad Stine, now the president of Buckhead Beef.

"In my lifetime, I have known a few guys who have impacted me and been a mentor," Stine says. "My father, grandfather, and Mick Colvin."

Left: Fred Johnson inducted in 1999. Courtesy of Kentucky State Fair Board and John Boyd Martin
Right: Mick Colvin inducted in 2009. Courtesy of Kentucky State Fair Board and Richard Halstead

Beef Legends

In Hollywood, the ultimate honor is a lifetime achievement award or star on the Walk of Fame. The Saddle & Sirloin Portrait Gallery, in Louisville, Kentucky, is the pinnacle for leaders in animal agriculture. Johnson and Colvin entered these hallowed halls in 1999 and 2009, respectively.

The gallery is one of the largest portrait collections devoted to a single industry. The earliest depict European stockmen of the eighteenth century, joined by those of American pioneers, all on display at the Union Stock Yards in Chicago from 1903 to 1977. That's when the gallery was relocated to the Kentucky Fair and Exposition Center. By tradition, the governing board selects one individual each year for a portrait.

SADDLE & SIRLOIN CLUB

Johnson's leadership extended beyond the brand to include roles as an Angus cattle breeder and Association Board member, first treasurer and second chairman of the National Beef Promotion and Research Board, and many others. While Colvin served the breed and industry in many ways, his success with the brand cemented his recognition.

"Mick Colvin steadfastly held to the belief that quality and consistency would result in consumer demand," says Bill Rishel, from Nebraska, a past chairman of the brand's board. "Through Mick, the brand's success created one of the great paradigm shifts in the beef industry as it ushered in the era of specification-based, value-added, brand-name beef and beef products."

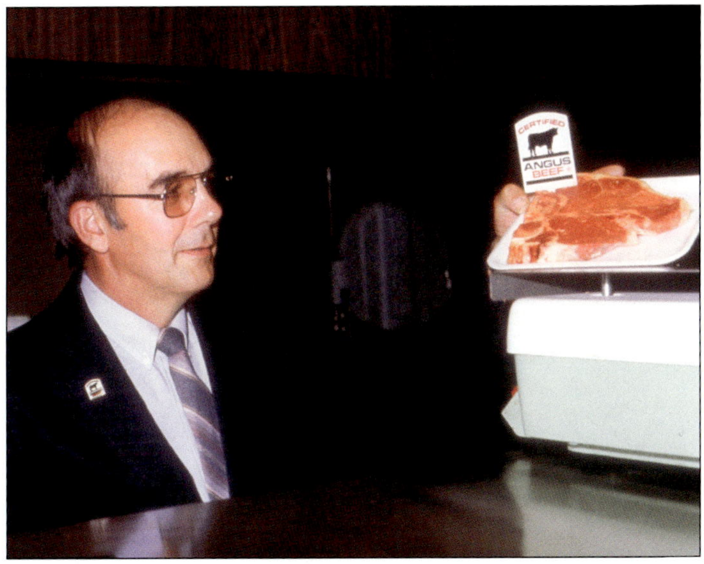

And the way Colvin announced his departure was as classy as the man himself. He told the leadership staff of his plans and then announced it to staff and licensees. But instead of stepping down in a month or six months, Colvin would take part in selecting his replacement. To make for a smooth transition, he would work with the new executive for one year before walking away. Essentially, Colvin announced his retirement in 1997 and actually retired in September 1999.

"It wasn't a surprise," Eichar says.

What was a surprise was how the interview process was handled. Because Colvin considered his employees to be partners, he insisted they would interview the candidates. The three people selected as finalists each had to give a presentation in front of the entire staff and then field questions. For many employees, it was a true honor to be a part of the process.

"Normally two or three executives make the decisions, but Mick always tried to keep the decision making around one table, like a family atmosphere," says Maggie O'Quinn. "We are family and we did make a lot of collective decisions."

For two decades, Mick Colvin is the face of Certified Angus Beef. He raises Angus cattle, speaks to young people and professionals alike, consults licensees, and wears many other hats to ensure the company lives up to its mission of improving demand for the Angus breed.

Colvin works alongside his replacement, Jim Riemann (left), for a year before handing over the reins to the program he knows inside and out. Colvin spends time introducing him to customers and sharing the passion and philosophy of what makes the brand so unique.

There was one candidate who seemed to have something in common with Colvin. And that was Jim Riemann, an executive from Excel packing. Riemann, a former Army infantry officer and Ph.D. meat scientist, taught at Kansas State and the University of Tennessee before joining Excel. He had the perfect pedigree for the job, and more.

"Jim stood out as the one of the three that had the most similarities to Mick's style of leadership, of really engaging people and turning them loose," Eichar says.

While Riemann might have been the best for the brand, Eichar says there was some concern whether he would move from Kansas to Ohio. The company offered him the position and to pay for his moving expenses, a $30,000 value.

Riemann was apprehensive about taking over a marketing company when his background was meat science. He prayed about the decision for three weeks and came to the conclusion: "There was only one way to find out if I could do this." He accepted, but declined money for moving. Instead, Riemann and his wife, Nancy, packed up their own belongings and made the trek themselves. For the next year, he worked alongside Colvin and learned about the company.

The transition was much easier than Riemann anticipated. "Sure, there was pressure," he says, "but I was very fortunate to work under Mick as associate director for a year. The thing I observed was not to micromanage people, but let people develop their own ways of doing things that enables them to grow professionally."

While everybody in the industry knew about the impact of the product, Riemann wasn't in awe of the branded middle meats and grinds. Rather, he was fascinated with the

Colvin respects his staff's attitude, pride, and commitment. They respect his integrity, leadership, and inspiration. They gather for a team photo in summer 1999, as his retirement nears.

People First

Among the brand's most avid supporters are the handful who can say they worked for Certified Angus Beef and are now with a licensee. As the brand progressed, some employees left for more lucrative opportunities that presented financial merits with which the not-for-profit company could not compete. However, that never ruffled feathers at the company. "I would rather have a good person for two days than not at all," Colvin says. "Three essentials for our success have always been the best people, the best product, and the best program."

The list of people who worked for Certified Angus Beef is remarkable. Many have moved on to roles with foodservice distributors, corporate retailers, packers, and the U.S. Meat Export Federation, for example. At their new places of business, they carry a voice that previously might not have been represented.

Chad Stine, former packing and retail director, says: "Certified Angus Beef is the envy of the branded beef industry and the envy of all programs out there. If there is a student looking for employment just after college in the meat or the food business, Certified Angus Beef is definitely a place you want to land. Part of that is because you are surrounded by great people."

During the year Jim Riemann (right) comes on board, he is like a sponge, absorbing everything he can from Colvin and staff. The duo regularly meets with executives, such as Brent Eichar (left) and Tracey Erickson, on the brand's course of direction. Photo credit: Sheryl Smith-Rodgers

"kitchen table" culture Colvin had created. When the company first started, the employees had meetings around Colvin's kitchen table and they were "so proud to be a part of that. I tried hard to maintain the kitchen-table culture and would always share the state of the company at staff get-togethers."

For the employees, Riemann's many similarities to Colvin eased the anxiety of replacing the seemingly irreplaceable.

Says Eichar: "Riemann is a man with lots of industry experience, but a comment that reflected his attitude was: 'I'm not here running the ship. I'm responsible for it, but this staff runs this business.' That was highly motivating to us because it was the same attitude that Mick brought. That philosophy, that attitude started with Mick and went right into the era with Jim."

Kind Goodbyes

As Colvin walked away, the letters poured in. Friends, former employees, licensees, and industry acquaintances, wrote to tell him of his value to the industry.

Some letters, like one from Franklin Hall, president of Lone Star Foodservice, talked about the writer's personal respect for Colvin.

"The one thing that I most respect about you is your deep belief and unwavering loyalty to the Certified Angus Beef program and its mission to increase demand for registered Angus bulls," Hall wrote on July 23, 1999. "Since we first met, your strategy and tactics were consistent with your articulated message, and you have stayed true to your word."

Everybody recognized his impact on the industry, but some talked about the future of Certified Angus Beef.

2003

Kansas State University Ag Research Center in Hays wins 2002 Best of the Breed contest; prizes for participants exceed $340,000

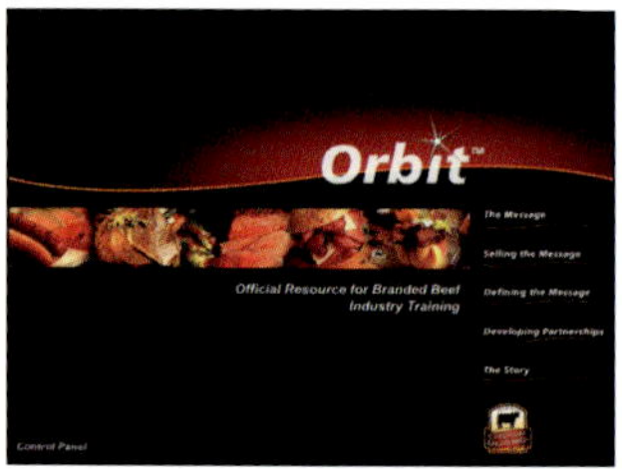

Official Resource for Branded Beef Industry Training (ORBIT) helps foodservice distributor representatives increase brand knowledge and confidence to sell it

National Angus Carcass Challenge begins four-year run; Mason Fleenor of GG Genetics owns champion pen of cattle

"Mouthwatering" campaign and "Angus beef at its best" tagline debut in target markets

Top: In one of his final messages to licensed partners, Colvin speaks at the 1999 annual conference about being "committed to the brand." He receives many heartfelt goodbyes from people who truly care about him and appreciate the brand's contributions to their livelihoods.

Below: Staff, customers, and producers give Colvin a fond farewell. At his retirement party, Colvin salutes his friends and colleagues. He also embraces his bright-eyed granddaughter, Caitlyn (left), and looks forward to more time for family, friends, and of course, golf.

2003

Licensed well-known chain restaurants in Mexico include Hard Rock Café and Planet Hollywood

More than two million carcasses certified during year

Products available at more than 5,500 restaurants and 5,200 grocery stores

U.S. case of Bovine Spongiform Encephalopathy closes doors to export markets

Trade-level "Picky, picky, picky" advertisement positions brand as beef category leader

Packers paid producers more than $30 million in premiums during year for qualifying cattle

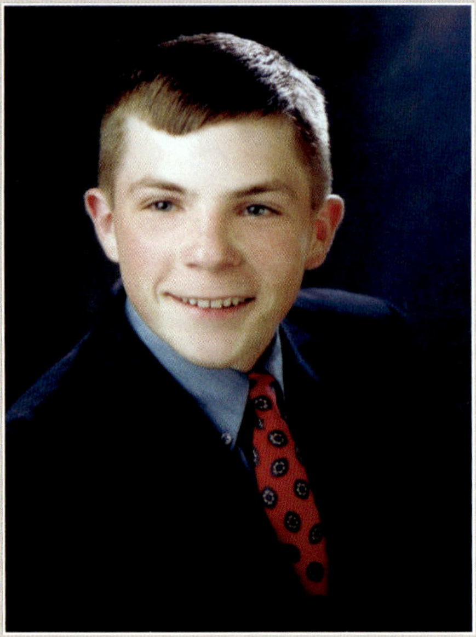

Ryan Sweeney, Mason, Michigan, receives the first Colvin Scholarship in 2002. A Michigan State University student, Sweeney has served on the National Junior Angus Association Board.

Scholarship

In honor of Colvin's service, Certified Angus Beef created a scholarship in his name. The Colvin Scholarship recognizes his role in making dreams reality and inspiring others to be their best.

College juniors and seniors who plan to be future beef industry leaders are eligible to apply. Every year, applicants are required to write an essay, but the subject changes from year to year. Since fundraising began in 1999, a total of $51,000 has been presented to 27 students.

G&C Food Distributors/Palmer Food Services, Rochester, New York, sponsored the Colvin golf tournament in 2008 and 2009. It and an auction at annual conference are the primary scholarship fundraisers.

"After being involved with the Certified Angus Beef brand for many years, we saw an opportunity for our companies to become involved more actively in an organization that has contributed to our success and provided greater visibility and prestige within our industry," said David Lepage, G&C president, in a news release. "It's exciting and rewarding for us to see young people who are motivated and instilled with a work ethic that is not all that common these days, and we need to do all that we can to support that. It's important to give back to the community and the industry that has provided all of us with a livelihood."

Brand partners from every segment of the industry gather in Wooster, Ohio, to salute Mick in 1999 as he retires. They establish the Colvin Scholarship Fund which included a donation from Tyson Fresh Meats. Representing Tyson are Rob Shuey (left), Kevin Munn (center), and Jerry Holbrook (fourth from left). Also pictured are Edd Hendee (second from left), from Taste of Texas, and staff member Chad Stine.

Mick Colvin often refers to early Certified Angus Beef licensees as pioneers of something great. During the brand's twenty-fifth anniversary at the 2003 annual conference in Cleveland, some of those pioneers—who are today great friends—gather for a photo. They are (from left to right): Eliott and Regina Reyes, Virginia and Mick Colvin, Sue and Bob VanStavern, June and Charlie Brown, and Norma and Arnie Hein.

"Your vision for branded beef was unprecedented," wrote Dick Bond, then chief executive officer of IBP, on September 17, 1999. "IBP is proud to be a partner in the CAB program and we look forward to working with Jim Riemann and his staff to build upon the foundation you helped create."

Others simply wanted to thank Colvin for making their business better.

"Due to the strict quality controls put in place by your organization, Certified Angus Beef keeps our focus to deliver this caliber of beef to our customers," wrote Joseph Dermastja and Ronald Chapchuk from Ontario's J.J. Derma Meats.

USDA's Michael Dunn wrote that Colvin's "hard work, dedication, and foresight have been at the forefront of the development and growth of a 'high quality' beef program that is the envy of dozens of other programs."

The staff hosted a retirement party for Colvin, in which they showed a special video comprised of interviews with Fred Johnson, C.K. Allen, Dr. Bob VanStavern, and various staff members. The documentary was a historical perspective of Certified Angus Beef and a look at how Colvin shaped the company.

"I first met Mick when he was manager of Mole's Hill Farm. They had four international grand champions—more than any place in the world," said Johnson in the documentary. "Because of his knowledge of the Angus business and determination and tenacity, we thought he would fight through any problems and stick with the program. We were right."

2004

First corporate chef, Dianna Stoffer, leads consumer and chef education

"Never, never, never" natural product launches with B3R Country Meats and Newport Meat Company

First restaurant licensed for natural steak: Mark's American Cuisine

As exclusive purveyor for American Express Celebrity Chef Tour, brand helps promote culinary arts in nearly 20 major cities

First foodservice distributor in Central America: Alimentos Campeon S.A.

Product success in Mexico leads Hard Rock Café to license for its Legendary Burger in U.S. locations

Mick and Virginia Colvin stand outside their farmstead in West Salem, Ohio. Mick calls Virginia the first "unofficial" Certified Angus Beef employee, and she certainly supports him throughout his tenure as executive director.

Photo credit: Sheryl Smith-Rodgers

The twenty-fifth anniversary celebration at the Cleveland conference is Jim Riemann's time to shine. During his first four years as company president, he leads the brand to unprecedented growth, marking a record 583 million pounds sold in 2003.

13 The Riemann Years

After he took the lead, Jim Riemann would say he was still riding with "training wheels" for about three years. But the truth was Riemann began taking Certified Angus Beef to the next level right away.

One of his first priorities was making Certified Angus Beef a limited liability company (LLC). With the Sysco case still fresh in people's minds, Riemann knew the company had to further protect itself from future litigation. In the restaurant and retail business, licensees frequently sue manufacturers and Riemann knew the company could be vulnerable to such suits, whether they had merit or not. His biggest fear was that the American Angus Association, which was named as the primary defendant in the Sysco case, could have its assets seized if the brand landed on the wrong side of a suit. Under an LLC format, the Association would be protected against judgments and lawsuits against Certified Angus Beef, which maintained its own accounts and assets. "It was important to establish some kind of legal organization that reduced risk to the Association," Riemann says.

Echoing Colvin's earlier struggles as executive director, Riemann found his recommendation to become an LLC met with contentious debate, but the board passed it in 2001, nearly two years after being proposed. At the same time of the LLC restructure, Certified Angus Beef became owner of all the trademarks. And the office of executive director became president.

Bryce Schumann, who turned the Association into an LLC after becoming its CEO in 2008, says he's confident Riemann made the right decision. "The structure we have adequately protects the parent company from risk but allows for that tie to the Association," he says.

2005

Representatives from 20 countries attend Roundup (formerly Boot Camp), making it the largest delegation to date

Boneless short ribs from Hudson Meat Company gain entry to healthcare

"Black Ink Basics" seminars shared with producers in South Dakota and Montana; related technical bulletins cover calf health, yield grade, and percent Angus

Gary West Steak Strips among 12 favorites from Fancy Food Show featured by Sara Moulton, a Food Network chef, on *Good Morning America*

In 2004, B3R Country Meats introduces Certified Angus Beef brand Natural, helping licensed businesses reach consumers who prefer natural foods. (From left) Tim Hussman, Newport; Mary Lou Bradley, B3R; and Riemann launch the new line.

While the LLC campaign may be Riemann's most significant contribution to the company, some licensees will argue that he most helped their companies by introducing Prime and Natural product extensions. As more Angus brands and packers' premium beef hit the market, the two new product extensions would greatly help licensees differentiate in the marketplace. But only with proper support from the brand.

Like Colvin, Riemann built and continuously improved relationships with licensees. But Riemann knew he or his Wooster staff could not be everything to everybody all the time. That's why he initiated a regional sales network.

"We recognized the marketing effort we put out was based upon developing relationships with our different licensees. We are educators who help licensees understand

Natural production expands through the Beef Marketing Group and Tyson Fresh Meats in 2006. John Butler, of the Beef Marketing Group, speaks about the program during a news conference at the National Cattlemen's Beef Association annual meeting.

not only the brand and its specifications but what the traits mean to their customers," Riemann says.

As Riemann and the leadership team were pondering those ideas, the Internet world was kicking into full gear. More than ever, salesmen from all industries were relying on cell phones and email. As a whole, society was getting away from the "human touch" and people were more concerned with their text-messaging than with real-person conversation. Certified Angus Beef would not follow that trend.

"We needed that eyeball-to-eyeball contact with the licensees," Riemann says.

2005

First television commercial produced for retailer use

Supermercados Amigo, Puerto Rico, teams up with E & J Gallo Winery for wine and meat case cross-merchandising

First Caribbean locations to offer Prime steaks: Prime Steak House and Rare

Interactive retail training CD launches, with educational components for meat and store associates

When National Beef Packing announces it will offer Certified Angus Beef Prime in 2001, premium U.S. steakhouses, including the Bull and Bear in the Waldorf=Astoria, could not wait to offer the steaks on their menus. Under Riemann's leadership, the line expands into retail as well.

The issue was being debated when a pair of valuable employees, Rebecca Thomas and Amanda Wydner, were moving away from Wooster because their spouses had career opportunities elsewhere.

"We didn't want to lose Rebecca or Amanda," Riemann says. "It opened the door for us to create these executive account managers and we quickly saw, through them, just how beneficial this could be."

The account managers would become invaluable to the company's mission just by regularly going to a distributor giving what Riemann calls "atta boys." He says the support and encouragement helped keep stability among licensees.

"There is such a turnover of sales people in these foodservice distributorships," Riemann says. "You might be selling shoes today but tomorrow you may be selling Certified Angus Beef products. Our regional executive account managers are right there to help put together training programs and provide support for whoever needs it. That has proven to be such a beneficial part of the program."

During his tenure, Riemann led the company to sales of more than half a billon pounds in eight consecutive years, expanded value-added products, licensed packing

Luke Walenciak, son of staff members Clint and Deanna, sports one of Nancy Riemann's baby bibs.

First Lady's Bibs

The First Ladies of Certified Angus Beef have a tradition of making an impact on the brand. Like Virginia Colvin, Nancy Riemann helped her husband through the tough times and made things fun through her artistic endeavors.

Whenever an employee had a baby, Mrs. Riemann, a long-time seamstress, made the parents a cute baby bib that said "I can't wait to eat ..." followed by a Certified Angus Beef logo.

"She sent one to a CAB specialist in California when he and his wife had their first child," says Mark Polzer, vice president. "To this day, he's never forgotten that!"

Riemann leads the brand through the challenges of international market closures to U.S. beef in 2003. Prior to the ban, Japan was a critical market for the company. In 2001, representatives from Restaurant Gyuemon come to the U.S. to tour a packing plant and Summitcrest Farms, and meet with Riemann and Mick Colvin. Executives share their journey with customers on this placemat.

plants in Canada, brought on two key brand extensions, went to the Olympics, saw the brand through world market closings, and helped the company grow to 12,722 restaurants and retail outlets. By all accounts, he had an amazing presidency. He maintained the brand's integrity and took it to tremendous heights.

Riemann's retirement was announced October 9, 2006, and took effect November 15.

"The board is grateful for Jim's dedication and loyalty to Certified Angus Beef and for his commitment to the mission, which is to increase demand for registered Angus cattle," said the company's board chairman Bob Norton in the news release.

While retirement had been on the horizon for Riemann, he cited pending changes in the brand's product specifications as a contributing factor in the timing. In a move to improve product consistency, the Certified Angus Beef Board of directors had recently voted to adopt new requirements for ribeye size, carcass weight, and trim not specifically addressed by the brand's yield grade standard.

"Changes in the overall beef industry in the many years since creation of the brand simply called for this logical response to better serve our customers," Norton said in the release.

"Jim did not agree with the direction we were going," says John Stika, current president. "He was very open about that with everybody. But it was never personal."

The change would not affect the two most important specifications, but it did have an impact on the size of animals accepted and it added uniformity to Certified Angus Beef cuts. Perhaps 6 percent of cattle previously accepted would no longer qualify based on their lack of uniformity. At the same time, an even greater percentage of cattle—that previously did not qualify, due in part to excess fat cover that is ultimately trimmed away by retailers and distributors—would be accepted under the new rules. Riemann was concerned that the change would encourage producers and feedlots to make their cattle too fat, and he still says, "Many of those Angus-type cattle were not genetically designed to be fed to such a heavy weight." Riemann believed brand standards should not be changed to include those cattle.

Most of the board—the majority of whom came from producer and meat science backgrounds—and everyone else on the executive team disagreed with Riemann's stance.

"Jim was perceived to be more than just a boss. He was a father figure for many—just because of his nature, his approach. When he decided to step down, that inherently brought up a lot of questions," Stika says.

Those questions were met with integrity. There was no mudslinging. Nonetheless, there was still "a little bit of fear he would leave under not-so-good terms," Eichar says.

But Eichar knew Riemann well enough to know he was a man of character and that he would leave with the brand in mind.

"Even if there was friction that existed, he handled himself in the most professional, in the most positive way, for the value of the brand," Eichar says. "I may have disagreed with him fundamentally, but I could never disagree with the attitude that he brought to that decision."

Eichar would become the interim president. On the day Eichar assumed duties, Riemann pulled him aside and said, "I support you, Brent."

"Jim was quick not to make it about himself," Eichar says, "but to make it about the brand and the company first."

For forty-five days, Eichar held the office of president. On the last day, he handed the office keys to his friend and long-time colleague, John Stika. "I had the opportunity to be a part of that interview process and ultimately cast a vote in favor of him being hired as president. In a unique twist of fate, I hired my boss."

As for Eichar's former boss, Riemann still loves Certified Angus Beef and is still revered by licensees for making the company more accessible and for listening to them. And Riemann? He misses the meetings, strategies, and most of all, the people.

"I miss the interaction with the people. There is such huge value in being inside these walls, and being able to say, 'Wow, that's a great idea. Is this something we could add to

Right: One reward of the brand's Olympic supplieship is the camaraderie. Riemann recalls how the brand's guests came together as strangers yet develop a kinship that lasts long afterwards.

Below: Menu items also include nachos and chili.

Larry Closs (left), meat director for Overwaitea retailer in Canada, and Riemann stand amidst a sea of signs and people at the 2002 Olympic Winter Games. While he is president, Riemann licenses Canadian packers, which fulfills growing demand in the market.

Riemann (left) and Heidi Scheffler, brand representative in the Northwest, hang with the Snowshoe Hare, a Salt Lake Games mascot.

Throughout his tenure, Riemann places much emphasis on value-added products. In February 2000, reporters interview him when the brand's suppliership of the 2002 Winter Olympics is announced. The year following the Olympics, value-added product sales grow 21 percent to 12.7 million pounds.

Fritz Usinger (left), of Milwaukee-based Usinger's, talks with local news celebrity Johnny Biscuit about a rising star at the games: Certified Angus Beef frankfurters.

Jim Riemann listens to a meat presentation at a "Beyond the Basics" seminar. Customers and staff appreciate Riemann's humble, attentive nature. He is always part of the team and easily approachable.

it, to make it even better?' Those kinds of interactions are just really stimulating and keep the adrenaline flowing," Riemann says. "I loved being able to send out a little note once in a while to somebody that said, 'I heard that you did this. Congratulations and whatever we can do to help you be even more successful, let us know.' "

2006

Hard Rock Café burger ad appears during Fox's "Best Damned Sports Show" leading up to Super Bowl

Natural product line expands with partners Tyson Fresh Meats and Beef Marketing Group's network of feedlots

During March and April, more than 1,000 stores participate in cross promotion with Louis M. Martini wines

Pioneer Lodge, Willow, Alaska, becomes northern-most licensed steakhouse

Research paper, "Declining Quality Grades: A Review of Factors Reducing Marbling Deposition in Beef Cattle," released and downloaded 1,400 times within a month from *CABpartners.com*

Above: With Riemann, employees always feel like part of the family. They enjoy the moment Riemann meets his first grandchild, Annika, at the 2003 annual conference. His daughter, Kristen, and son-in-law, Justin Wiebers, travel from Tennessee to Ohio to make the introduction.

After retirement in 2006, Jim and his wife, Nancy, journey around the country in their motor home, a long-time dream. They relocate to Kansas to enjoy time with family and friends.

Photo credit: Sheryl Smith-Rodgers

2006

"The Brand That Pays" slogan registered for supply development promotions, and "Targeting the Brand" mark available for producer use at sales and events

AngusSource becomes genotypic option for cattle identification

Trademark registered in Madrid Protocol, representing almost 60 countries

Dr. John F. Stika named president upon Jim Riemann's retirement

In 2008, the Association enjoys 125 years of helping Angus producers, while Certified Angus Beef turns thirty. Association president Bryce Schumann (left) joins Mick Colvin (center) and John Stika to toast the future during the brand's annual conference in Coeur d'Alene, Idaho.

14 The Future

On Mick Colvin's last day as executive director for Certified Angus Beef, he spoke to his staff affectionately, in words so moving that employees still remember them. "I just want to congratulate everybody for becoming the number two branded fresh beef program in the world," he said.

Everybody in the room looked at each other, scratching their heads, as Colvin kept talking. Then, he repeated, "We've become the number two branded fresh program." Once again, his employees gave each other glances and then somebody stood up and said, "Uh, Mick, we're the number one program in the world."

Colvin smiled. "That's right. And the minute you start acting like you're the number one program, somebody will pass you."

In 1999, Colvin essentially laid out the challenges for the next twenty years. The marketplace has certainly changed since his tenure, but the brand's status as the number one branded fresh beef program has not. "As long as we always have integrity, the company will be fine," he says.

For president John Stika, who has worked solely for Certified Angus Beef in his career, everything about the brand revolves around what Colvin calls the three pillars: product, people, and integrity.

"Those three things are not unique to Certified Angus Beef, but the consistency in which they have been applied over thirty years is what has made this brand successful," Stika says. "It's the unwavering attention to those three things that Mick put in place, that Jim reiterated and further solidified. My job has been easy: to further propagate what they have made successful."

Above: President John Stika, only the third president for Certified Angus Beef, begins his career with the company in 1999 in the feeder-packer relations division. With a doctorate in meat science and strong ties to the cattle industry, Stika is constantly searching for opportunities to grow the brand's relevance for all, from producer to chef and retailer.

Right: Built in 1994, the company's headquarters is a place where coworkers roll up their sleeves and work smart and hard as a team. There's plenty of fun, good food, and a commitment to excellence. The company is known as one of the best places to work in northeastern Ohio.

The Culture

In the beef industry, Stika says, "The people at Certified Angus Beef are considered the gladiators, the protectors of quality. That carries a sense of pride."

Even Stika's competition appreciates his company. Certified Hereford Beef CEO Craig Huffhines says the Angus brand has greatly benefited the beef industry and made it easy for other brands to start up. Granted, he enjoys competing with Certified Angus Beef and takes pride in winning accounts, but has nothing negative to say about his Ohio-based competition.

"I have the greatest admiration for the Certified Angus Beef program," Huffhines says.

That kind of admiration is not just in the beef industry; it's in the company's hometown, too. When Kyle Miller, executive account manager for the Great Lakes region, interviewed for the job, his wife, Johnna, decided to have her nails done. As the beautician and Johnna small-talked, Kyle's interview came up. The woman said, "Wow, that is a really great place to work."

There are many similar stories throughout the town of twenty-five thousand. People know about the company and when there's a job opening, hundreds apply. "When you hear points like that, you know people feel valued and are in a positive culture," Stika says.

That culture has been tested numerous times: the Sysco litigation; Riemann's sudden retirement; and the passing of Marylynn Roe, a beloved employee whose husband, Bill, also worked there. Stika says her death from cancer in 2006 was the single most difficult challenge the company faced in a decade.

"She was one of the most pleasant people I've ever met—she had a great demeanor with people. Her death was tough on everybody," he says.

Certified Angus Beef staff believes in its mission of helping Angus producers, customers, and consumers alike. Most of its one-hundred-plus staff is based in Wooster, Ohio. The company has a small Manhattan, Kansas, office focused on supply development, as well as a dozen regional staff working from home offices in Japan, Mexico, and throughout the United States.

The "employee-first" attitude leads to some decisions that are not performance based or even financially driven. Rather, they're made with the thought of maintaining the culture and employee morale.

"The core item we sell isn't tangible. We sell a concept. We sell an attitude. We sell a philosophy. We sell an emotion about quality. I can't imagine trying to do that in an environment where people didn't feel valued," Stika says.

Meanwhile, the competition, much of which comes from licensed packers, has created premium brands that closely rival Certified Angus Beef. Stika believes the company cannot move forward by only thinking about how to react to new brands. "You just try to educate, educate, educate to get everybody talking accurately about what the true differences are, and let the consumer decide. That is part of our culture," he says. "We are charged with promoting the heck out of Certified Angus Beef while not disparaging other programs, which is hard to do in some cases. But you focus on this brand, what this brand does, not what other brands aren't doing."

As discussed in Chapter 5, some Angus breeders are in favor of the "me-too" Angus programs. They contend that Certified Angus Beef needs to create a lesser-quality line—possibly even a Select program—to capture the low-price-driven customer. But Stika says the program always comes down to the mission of driving demand for registered Angus cattle through a science-based product with superior taste.

The brand's success has elevated the Angus beef category in the minds of consumers and chefs, but not all Angus beef meets Certified Angus Beef's high standards. Direct campaigns and educational outreach focus on setting the record straight.

"To say that we should support other Angus brands because they drive demand for Angus cattle is assuming too much. For example, we won't support an Angus-cow beef program," Stika says. "That does not add value to the Angus breed or bring out the best in what Angus cattle can offer consumers. ... Alpo offers dog food made with Angus beef. Is that good for the breed?"

Speaking of dog food, Certified Angus Beef has been approached numerous times by pet food brands to market chews and dog foods. "If it's not for human consumption, we're not interested," he says.

As for new products, value-added takes center stage. Stika would like to see more pre-cooked entrées such as pot roast, short ribs, and deli meats, in stores and restaurants.

"We've got to continue to find ways to drive carcass utilization, to drive the packers' margin-per-head basis and thus, the premium to producers. If we're successful at that, we will have greater access to more product in the box than what we see today," he says.

Consumers' Marbling Preference

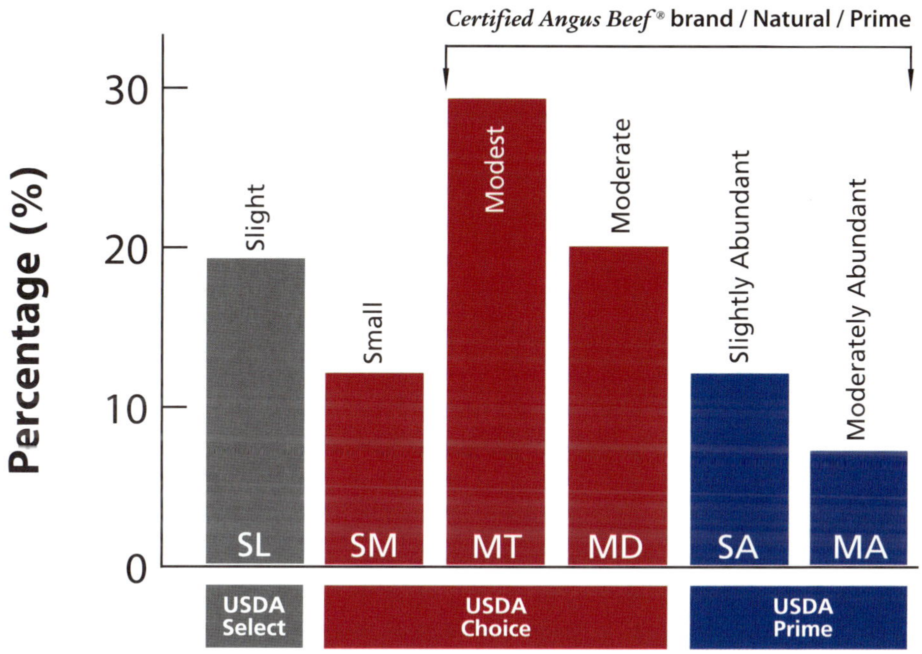

Research in 2009 out of Colorado State University points to "Modest or higher" marbling as the threshold for great-tasting beef. "If beef tastes great, people are not only more likely to buy it but also more willing to pay more for it," said Dr. Daryl Tatum, a meat scientist.

In the retail and foodservice sectors, there is a refocus on middle meats, like strips, ribeyes, and tenderloin steaks. With the expansion of ground beef availability and single-item foodservice licensing, some licensees have opted only for ground beef.

Stika envisions ground beef being an important add-on sale but not the primary buy for restaurants. He wants to maintain the tradition that, "Certified Angus Beef is the crème de la crème of middle-meat steak items."

Supply Lines

The future of Angus genetics looks bright. With carcass data from feedlots, DNA, and ultrasound technology giving breeders more tools than ever, seedstock producers are shooting for as high as 80-percent Certified Angus Beef acceptance rates from their sires. How can the brand take advantage of those trends?

2007

Specification limits for ribeye area, carcass weight, and external fat further ensure consistent sizing

Buckhorn Steak & Roadhouse wins National Beef Backer Award

MasterCard's popular "Priceless" commercial includes brand's ground beef

New chuck cuts launched with NCBA at Meijer stores in Cincinnati

Partnership with "Trump Steaks" gains national exposure at Sharper Image stores and on QVC

June sales largest ever at 56 million pounds

Supermercados Nacional promotions and media events introduce brand in Dominican Republic

"Expert" consumer campaign launched in target markets Albany, New York, and Wooster, Ohio

Educational programs get down to the basics, down on the farm, or in this case at Beller Feedlot for Roundup 2009. Building customer loyalty to the brand sometimes requires seeing to believe. This grassroots effort, much like waitstaff training, forms a network of support for the brand throughout the country and world.

2007

"Start to finish" ads for trade publications focus on brand's connections with Angus cattle ranchers

Best Practices Manual published as a cow-calf guide to producing cattle for the brand

Consumer research: 87 percent recognize brand name, 91 percent logo; that's 3.5 times more than closest ranking other brand

CattleFax estimates Certified Angus Beef brand and other premium USDA Choice and Prime programs add about $500 billion to cattle industry each year, or an extra $19 for every head of finished cattle

Stika says, "Our supply development team will continue to maintain and grow our position and the level of respect in the industry. I'd like to see us continue as the most tunnel-focused Extension program you could ever find." That reference is to USDA's nationwide Cooperative State Research and Extension Education Service, which must maintain a broad focus.

Schumann believes the American Angus Association as well as Certified Angus Beef must continue to provide quantifiable information to producers.

"It's the information age, but the level of that flowing through the cattle-production system is limited. We are really working to change that," Schumann says.

Corah calls it "economic documentation," and he says communicating to the producer is a pivotal focus for the future. The vice president for supply development says his division will continue sponsoring major producer gatherings, providing timely radio sound bites to farm shows, and news releases to ag media.

As far as the messages in those releases, the information will aim to maximize producer profits using Angus genetics with a focus on the Certified Angus Beef brand target. It's the company's hope that such constant communication will result in greater quantity of quality.

"We need to have an indirect impact on the industry that will find us another million carcasses annually in the next twelve years," Stika says. "We need to grow what we certify from 2.5 million to 3.5 million carcasses per year by 2020. And that onus is on our supply development team and our ability to create pull-through demand at the consumer level."

Top: Supply development staff work alongside the American Angus Association to form "Team Angus." Gary Fike (left) and his colleagues build solid relationships with producers and feedlot managers. The team helps them improve carcass value and increase profits.

Left: Improving the breed improves the brand. Dedication to delivering exceptional beef for consumers begins on the family Angus ranch. That's where the decisions are made that lead to calves with more genetic potential, managed with care, to profitably satisfy consumer demand.

2008

The Rex, Billings, Montana, wins National Beef Backer Award

Cooking with Kids contests and "create your own spice activity" in target markets

Midan Marketing and Shugoll study shows more than 58 percent buy Certified Angus Beef brand or know about it; consumers also give it highest rating among brands, earning 8.5 on 1-to-10 scale

Foodland in Hawaii introduces brand on four islands in 30 stores

King's Command Foods introduces branded meatloaf for deli rotisseries

2008

Chappell Feedlot hits gold level (2,000 head cumulative) in 30.06 program; awarded engraved 30.06 rifle

Cooking videos for braising, sautéing, and grilling debut on *YouTube.com*

In-store promotions held with Propane Education Resource Council, E & J Gallo Winery, and Lea & Perrins Worcestershire sauce

"Here's the Premium" study reports $34.51 per head advantage for Angus steer calves over other breeds—more than double $13.76 from fall 1999

Dinners with media and Angus producers held during "Great Steak Celebration" at James Beard House and Bull and Bear in Waldorf=Astoria, marking the brand's 30th anniversary

During first 10 years of feedlot licensing, data recorded on 1,370,980 cattle

Grid premiums paid by packers for qualifying cattle during past 12 years exceed $250 million

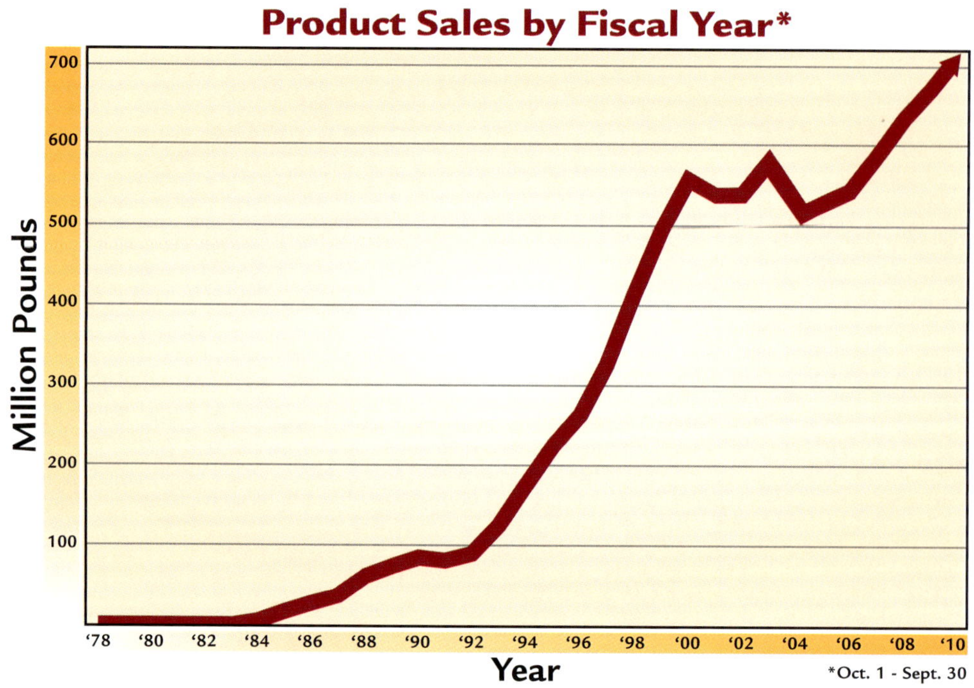

In the coming years, supply development will spend more time understanding the dynamics of economics and how they impact genetics, stocker operations, and the feedlot sector. "We need to find better ways for everyone to be profitable producing a high-quality product," Stika says. "It's a big responsibility that those guys have."

He sees a burgeoning export trade sector in the next ten years, with international sales potentially reaching about 25 percent of the company's total business. "When you look at where wealth accumulation is starting to pop up around the globe, there comes the desire for higher quality protein," he says. "There will be huge opportunities for us globally. Every protein company here in the States is starting to understand the value of the global market."

In the late 2000s, the company was selling about seventy million pounds worldwide, annually. Stika would like to see that number become around two hundred and fifty million pounds in the next decade. By 2020, he also wants to see a billion pounds sold overall from those 3.5 million cattle per year. Those are attainable goals if you consider the company's history and impact on the cattle business.

When asked in a 1999 interview if he, in the early days, fathomed Certified Angus Beef selling one billion pounds a year, co-founder Fred Johnson just grinned. "My belief was we might get to five to ten million pounds, but never gave it much thought past that," Johnson said. "Our purpose in the beginning was to create a strong demand for Angus bulls. We've done that."

More than a decade after Johnson's on-camera interview, Certified Angus Beef continues to deliver for the consumer and producers. One of them has been Jerry Helgren, who has watched this program from the very beginning as a participant in the early meetings and as an Angus producer at Picket Fence Farm in Illinois.

As company executive Mick Colvin states from earlier times, "We always want to be connected to the farm." Telling the rancher's story and their dedication to quality helps consumers see the unique ties this not-for-profit brand has to America's family farms.

2009

Charlie Brown's restaurant chain wins National Beef Backer Award

Foodmaster Supermarkets wins NCBA retail award for chains with up to 10 stores

New York City-based media, including *New York Times*, *New York Post*, *Food Arts*, and *American Heritage*, gather at Trowbridge Angus for "A Day at the Ranch"

Satellite media campaign with safe grilling tips held with celebrity chef Rick Bayless

Chef John Doherty helps promote at Share Our Strength Championship BBQ during National Restaurant Association Show

Reasor's, Oklahoma, becomes first retailer to offer Certified Angus Beef brand Prime Natural

2009

Comercial NorteAmericana (ComNor) launches Prime and Natural in Mexico

First Canadian restaurant for Prime: La Bête bar-steakhouse

New advertising tools help foodservice and grocery licensees connect with Angus producers in every state

Sales exceed 60 million pounds monthly from July through October; annual record at 663 million pounds

Acceptance rates climb to 19.8 percent, continuing a four-year trend

More than 500 new international restaurants sign on; new distributors in Bahrain, Hong Kong, Japan, South Korea, and Suriname

Nearly 7,000 menus collected and 64,200 audits conducted for restaurants and grocery stores during year

CertAngusBeef on Twitter launched; fans also follow brand on Facebook

Certified Angus Beef is growing new markets, reaching new consumers, and satisfying the craving for great-tasting beef.

"Certified Angus Beef is probably the best thing that came along to help promote Angus cattle," Helgren says. "It's been tremendous."

The program has certainly set tremendous volume marks for pounds and dollars. In 2008–09, when the U.S. and world economies were tanking, Certified Angus Beef licensees sold a record 663 million pounds, more than twenty million pounds above the previous year. Those sales generated about $2.7 billion from consumer spending. A company survey of packers indicated they had paid out more than $280 million in premium bonuses in ten years' time to producers of cattle that hit the Certified Angus Beef brand target. From the outside looking in, one might say Certified Angus Beef has achieved its mission and could easily sail into the future. But for Colvin, who spent a big part of his life building the brand, Certified Angus Beef is still second best—compared to its future standing.

Even though the company has become a larger-than-life organization for Angus producers and the U.S. herd is mostly black; even though licensees have made millions selling the brand, and consumers recognize value in the brand, the retired Colvin still isn't satisfied. He sees so much more potential in Certified Angus Beef. He simply asks: "What's next?"

There are opportunities for the brand to help customers market all beef cuts, but the bread and butter of its business will always be classic, grilled steaks. Leading with ribeye, strip loin, filet mignon, and top sirloin draws attention to Certified Angus Beef's unmatched dedication to quality, which ultimately wins over chefs and consumers.